# UNDERGROUND COAL GASIFICATION

# UNDERGROUND COAL GASIFICATION

George H. ~~Lamb~~ Greus

NOYES DATA CORPORATION

Park Ridge, New Jersey, U.S.A.

1977

CHEMISTRY

6231-2005 ✓

Published in the United States of America by
Noyes Data Corporation
Noyes Building, Park Ridge, New Jersey 07656

# Foreword

This is the fourteenth volume of our Energy Technology Review series.

As is the case with several other volumes in this series, most of the information presented in this book is based on federally funded studies. Here are collated and condensed vital data that are scattered and difficult to pull together. Experimental equipment and set-ups are reviewed and detailed by actual case histories. Special attention is paid to advance indications deduced from the present energy crisis.

Advanced composition and production methods developed by Noyes Data are employed to bring our new durably bound books to you in a minimum of time. Special techniques are used to close the gap between "manuscript" and "completed book." Technological progress is so rapid that time-honored, conventional typesetting, binding and shipping methods are no longer suitable. We have bypassed the delays in the conventional book publishing cycle and provide the user with an effective and convenient means of reviewing up-to-date information in depth.

The table of contents is organized in such a way as to serve as a subject index and provides easy access to the information contained in this book. Most chapters are followed by a list of references giving further details on these timely topics. The bibliography at the end of the volume lists the highly important government reports and a source of their purchase.

3072

# Contents and Subject Index

# Introduction

Until recently, the United States used its energy resources as if they were essentially inexhaustible. The need for concern over depletion of these resources seemed far in the future. The present, however, is now the future and the days of plentiful supplies of inexpensive energy have become the past. Unfamiliar terms—fuel shortage, energy crisis, resource allocation, rationing—are now familiar to nearly everyone. Any breakdown in our national energy supply is a major crisis. Past progress in America—in employment, standard of living, welfare of society—was due in large part to the abundant supplies of cheap energy that were available. Without sufficient energy sources, the nation's well-being internationally and domestically will be adversely affected, resulting in greater material and manpower costs to maintain our standard of living.

The role of oil and natural gas as sources of energy is almost as well known as the shortage of these vital fuels. As recently as 1972 more than three-fourths of the energy used in the United States was derived from these sources. Coal, one of the most abundant mineral fuels, supplied only 17% of the energy needs that year.

In view of the looming shortages of oil and natural gas, all possibilities of energy production must be explored. With the large United States coal reserves, special interest has been attracted by the underground gasification of coal deposits. Deposits that could not be mined competitively by other processes might be economicaly utilized by underground gasification. In general, the coal is gasified in situ and the gas produced is collected, conveyed to the surface and utilized further. By underground gasification of coal reserves, the energy content of these deposits is obtained in the form of gas and heat.

With the passage of The Federal Coal Mine Health and Safety Act (Public Law 91-173) in 1969, emphasis was placed on safer mining practices to

1

minimize the occupational hazards associated with coal mining. In situ coal gasification therefore became attractive since it could extract the energy from the coal through surface boreholes, minimizing the health and safety problems associated with deep mining. Another advantage of successful underground gasification includes formation of a fuel gas that may be cheaper than other energy forms. From an environmental standpoint, hydrogen sulfide rather than sulfur dioxide is the predominate form of sulfur produced in underground coal gasification. The small amounts of hydrogen sulfide produced can be treated by hot carbonate scrubbing, but no effective commercial-scale means for removal of sulfur dioxide exists.

This book presents an overall view of underground coal gasification starting with its history, and comprising the theoretical basis for the method, the typical processes, including U.S. program studies, the problems encountered, special instrumentation required, foreign advances, economic aspects and future outlook.

Although in situ gasification of coal may not eliminate all problems, it appears to have great promise in alleviating some of the problems associated with production, transportation, and consumption of this fuel.

While not expected to replace coal mining, it offers an alternative method of bringing the energy value of the coal to the surface. If feasible, this clean, low-Btu energy form could supplement the gas supply, particularly in the industrial sector where it could be used to produce electricity.

# General Background and History

The material in this chapter has been excerpted from CONF 75-1171-1, PB 299 274, PB 241 892, PB 256 155, UCRL-51217 and UCRL-Trans-10866. For a complete bibliography, see p 251.

## ENERGY REQUIREMENTS

The United States is confronted with a growing deficit in energy of all kinds. The electrical energy deficit is projected to be overcome with additional nuclear power plants followed in several decades by electrical energy from controlled thermonuclear reactors and from geothermal (1) and solar (2) energy sources. Among the fossil fuels that now supply the majority of our energy needs, shortages in both gas and liquid petroleum are imminent while coal is in abundant supply (3).

Since 95% of the energy used in transportation comes from liquid petroleum (4), projections are that future requirements must be met by increased imports, since domestic production cannot meet this demand. The result is increasing deficits in balance of payments and an increasing dependence on the availability of import sources. The shortage of natural gas (mostly methane) is most severe. This gas is now used mostly for industrial purposes and household heating (4). Electrical energy can supply most of these requirements only if furnaces and appliances are changed. Further, if large resources were available, natural gas or liquid petroleum manufactured from methane could be used for transportation and would satisfy our pollution requirements without major modifications in engine or exhaust systems (5).

The obvious combination of technologies that would contribute to solving both environmental problems and energy shortages in the near and intermediate future, especially for gas uses, is to convert a portion of the ex-

3

tensive coal deposits to gas. This solution was highlighted in the Presidential Energy Message to Congress in June 1971 in which the third major element was ". . . an expanded program to convert coal into a clean gaseous fuel." Several processes are now being studied for converting strip-mined coal into clean gaseous fuel (6)(7)(8). Numerous factors must be considered (requirements for water, environmental effects of strip mining and shipping the coal, removal of sulfur, pollution from fly ash, and waste disposal), but the cost of mining and the plant investment (9)(10) suggest the desirability of operating the coal-to-gas conversion process in the ground.

Coal deposits of one kind or another are known to exist in almost every state of this country, but by far the largest reserves are in the Rocky Mountains and Colorado Plateau regions. The most interesting and potentially useful coals for in situ processing are those between 600 and 3,000 ft below the ground surface.

It is interesting to note that 1 acre-foot of coal can yield as much as 40 billion Btu of thermal energy. There are some 100 abandoned coal mines on fire in the United States. The fact that some of these have been burning for well over 50 years suggests that an in situ coal combustion process is maintainable for extensive time periods.

Figure 1.1 shows areas in the Western U.S. in which coals of the appropriate chemical and physical characteristics are to be found. According to a U.S. Bureau of Mines summary of coal reserves, these deposits represent more than $1.5 \times 10^{12}$ tons (11). If it proves possible to process 30% of it, these coals represent the equivalent of about 10,000 trillion ft$^3$ of gas or 30 times the present gas reserves. Thus, the in situ gasification of coal could make a significant impact on the energy crisis, extending our fossil resources by perhaps as much as 100 years.

## HISTORICAL DEVELOPMENT

Attempts to produce gas by in situ processing techniques began in the mid-nineteenth century (12). As early as 1869 W. Simons (USA) suggested (13) gasification of ash-rich coal underground; in 1888 D. Mendeleev (USSR) (14) took up this thought again, but without adding any essentially new ideas. In 1909 A. G. Betts (USA) (15) was granted a British patent for a process of underground gasification based on the fact that the coal was gasified in its natural deposit by suitably arranged boreholes. Finally, in 1912 W. Ramsey (Great Britain) made the first preparations for a practical trial which, however, never came to pass because of the outbreak of the first world war.

**FIGURE 1.1: COAL FIELDS OF THE WESTERN UNITED STATES**

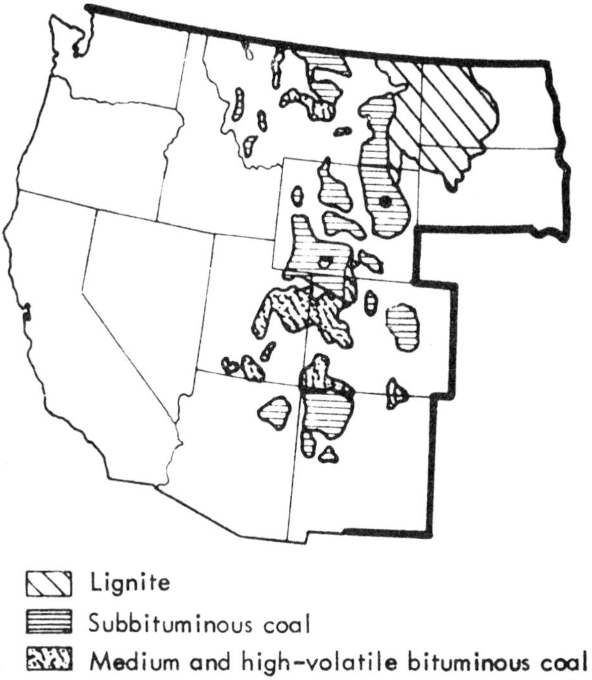
Lignite
Subbituminous coal
Medium and high-volatile bituminous coal

Source: UCRL-51217

Between 1914 and 1930 no essential research on underground gasification was carried out in the western world. In contrast to this, Lenin had underground gasification included in the Soviet research program during the first world war. Between 1928 and 1939 the first practical trials were prepared in the USSR. These trials were carried out by Podzemgaz, an organization run by the state, which established five test stations in all.

Thus, in 1928 the first experiments with hard coal were performed at Lisichansk (Donets basin) (16). In 1933 tests in the lignite area of Krutov (Moscow basin) followed; these were mainly aimed at establishing the practical basis and the conditions of application of underground gasification. The research work was unfortunately interrupted by the beginning of the second world war. After the war European and overseas countries again showed interest in underground gasification, which had been tested in the USSR and had partly proved useful. From 1947 to the beginning of the 1950s practical experiments were carried out in Italy, Great

Britain, France, Belgium, and the USA. Extensive research programs in Czechoslovakia between 1956 and 1966 led to further valuable results (17). In recent times the tendency toward further systematic investigations in underground gasification has risen worldwide. Most of the recent programs have been directed to shallow deposits of subbituminous coals in eastern Germany and in Russia around Moscow. Smaller programs also have been conducted in other nations including the USA (18).

## United States Involvement

Three reasons appear to have supported the resurgence of interest in the United States in the in situ gasification of coal. The first was the passage in 1969 of the Federal Coal Mine Health and Safety Act which placed an increased emphasis on safer mining practices to minimize the occupational hazards associated with underground coal mining. This law, at the time of passage, produced its larger impact in the eastern mining operations which are largely underground. Attention then had not yet begun to focus on the western coal seams that can largely be mined by open pit methods. Within two years from the time the law came into effect, by 1971, coal prices at the mine in carload lots had about doubled.

The second reason occurred later in 1973/74, when world petroleum prices rose markedly as a result of the joint price control actions instituted by the petroleum exporting nations, and a renewed importance was placed on the reliance on U.S. coal resources for supplying the domestic demand for electricity and natural gas. But, this reason applies equally well to interest in the gasification of coal by any means other than in situ.

The third reason has been the rapidly increasing public attention and legal force being given to the protection and preservation of the environment. Especially since 1970, differing adverse environmental impacts have been associated with both open pit and underground methods, and the expectations are that the problems of environmental protection associated with in situ gasification can be less severe.

## BENEFICIAL IMPLICATIONS

The concept of efficient thermal energy generation through in situ coal combustion combined with on site conversion of the heat energy to a transportable form (for example, electricity) would offer distinct economic and safety advantages in the utilization of our coal reserves, as follows:

[1] The amount of underground mining needed to extract energy

from coal can be greatly reduced. Thus coal utilization can be increased without a proportionate increase in mining personnel and equipment or in coal-handling and transportation facilities.

[2] Energy can be made available from coal deposits which are not economically minable by conventional techniques (for example, low-heat-value coals and thin seams).

[3] Energy can be made available from previously mined coal seams, some of which may contain as much as 50 percent of the original coal.

[4] In situ combustion might be used to supplement other in situ coal extraction processes, such as gasification and liquefaction, to optimize the recovery of the total energy of the coal.

It therefore follows that the three major differences between in situ gasification and conventional mining of coal are:

[1] In situ gasification might recover known reserves that cannot be mined by conventional techniques, e.g., steeply inclined seams and thick deep seams.

[2] In situ gasification might increase coal reserves by making it possible to treat coal seams excluded from the current U.S. reserve picture, for example, seams thinner than 14 inches or seams deeper than 6,000 feet.

[3] For deposits amenable to both in situ and conventional techniques, a higher recovery of the coal via in situ gacification may be possible.

## U.S. RESOURCE ASSESSMENT AND CHARACTERIZATION

It is a widely known fact that the higher the rank of coal, the poorer its in situ gasification quality. The higher the volatility and reactivity and ash content up to 50%, the better the coal for underground coal gasification. Thus the general order of priority on coal type in this country is lignite, subbituminous, bituminous and anthracite. However, it is believed that anthracite cannot be gasified in place to produce a satisfactory low Btu gas. Distribution of the coals in the U.S. is illustrated in Figure 1.2. The obvious dilemma is that the best candidates for underground coal gasification do not occur in the Eastern U.S. where the demand for gas is greatest. The alternatives then exist to [1] gasify the lignite or subbituminous coals and generate electricity on site for transmission to the demand centers, [2] upgrade the underground coal gasification low Btu gas to pipeline quality, [3] gasify some of the otherwise unrecoverable bituminous coalbeds or [4] control the underground coal gasification

**FIGURE 1.2: COAL FIELDS IN THE UNITED STATES**

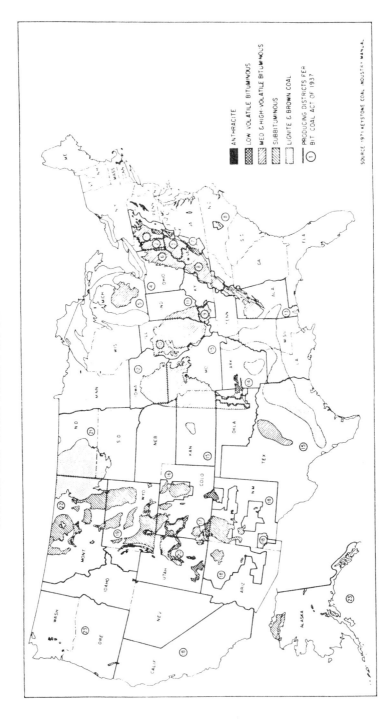

Source: PB 209 274

process and supplement injections with oxygen and steam to obtain a technological gas or feedstock for the production of ammonia, methanol or other products as described later and demonstrated in the Soviet Union. The greatest negative factors associated with electrical transmission from Western sites are the costs and security involved in the required distances which exceed 2,000 miles.

Another factor weighing heavily upon the decision to gasify a coal reserve in situ is whether it is recoverable by existing technology. Recent developments in mining technology (19) indicate that deep, thick coalbeds may become mineable in the near future.

The magnitude of resource availability must be such that regional and local development could extend for a period of at least 20 years. As illustrated in Figure 1.3 (20), sufficient potential resources exist in all regions of the U.S. to meet this requirement.

## FIGURE 1.3: POTENTIAL IN SITU RESOURCE

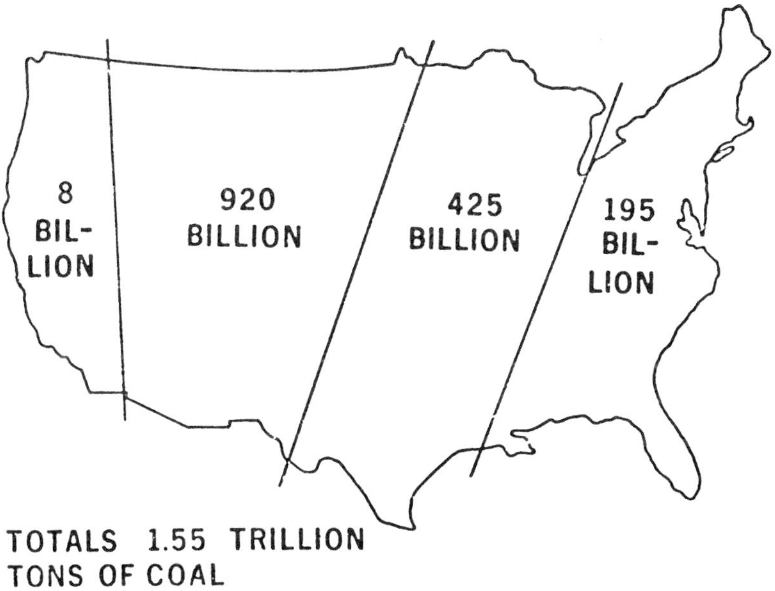

8 BIL-LION    920 BILLION    425 BILLION    195 BIL-LION

TOTALS 1.55 TRILLION TONS OF COAL

Source: CONF 75-1171-1

The fundamental properties of coal: heating value, type, reactivity, and volatility are essential to all the known underground coal gasification concepts. However, many other properties dictate the particular gasification scheme necessary for successful underground coal gasification. The more important of these properties are:

[1] coal permeability—frequency and distribution of fracture systems

[2] groundwater circulation—acquifer systems within, adjacent to or through the coalbed

[3] bed thickness

[4] number of and spacing between coalbeds

[5] depths from surface

[6] inclination of coalbeds

[7] regularity of bed thickness and slope

[8] over- and underlying strata characteristics including; porosity, gas and water permeability, thermal and mechanical properties, thickness, and mineral type

[9] coal thermal-mechanical properties

[10] directional thermal, mechanical and flow properties of coalbed

[11] various chemical properties

No evidence has been found that directional coal properties have been used in planning experimental or commercial underground coal gasification plants. Perhaps, other controlling factors, such as bed slope, usually prohibited such use. However, in many of the coalbeds in the U.S. conditions are such that the directional properties can be utilized to an advantage. Gasification at depths greater than 1,500 feet has generally been considered uneconomical due to the increasingly higher compression costs required initially with depth. To cope with these widely varying conditions, different schemes which are predicated upon envisioned concepts of the in situ process conditions and mechanisms have been considered and are described more fully hereinafter.

## REFERENCES

(1) E. R. Berman, *Geothermal Energy*, Park Ridge, N.J., Noyes Data Corp. (1975).

(2) A. R. Patton, *Solar Energy for Heating and Cooling of Buildings*, Park Ridge, N.J., Noyes Data Corp. (1976).

(3) M. K. Hubbert, *Energy Resources for Power Productions*, International Atomic Energy Agency, Vienna, Rept. IAEA-SM-146/1, presented at Symposium on Environmental Aspects of Nuclear Power Stations, New York (1970).

(4) Arthur D. Little, Inc., *Energy Policy Issues for the United States During the Seventies*, prepared for National Energy Forum (1971) p. 3.

(5) Stanford University School of Engineering, *Air Management Recommendation for the San Francisco Bay Area*, NASA Contract NGR-05-020-409, Stanford University (1971) p. 131.

(6) G. A. Mills, "Gas from Coal—Fuel of the Future," *Environmental Science and Technology* 5 (12), 1178 (1971).

(7) F. L. Robson et al., *Technological and Economic Feasibility of Advanced Power Cycles and Methods of Producing Nonpolluting Fuels for Utility Power Stations*, U.S. Dept. of Health, Education, and Welfare, Rept. PB 198 392 (1970).

(8) W. S. Doyle, *Strip Mining of Coal—Environmental Solutions*, Park Ridge, N.J., Noyes Data Corp. (1976).

(9) *U.S. Energy Outlook—An Initial Appraisal 1971–1985, Vol. 1*, National Petroleum Council, Washington, D.C. (1971).

(10) J. P. Henry, Jr. and B. M. Louks, "An Economic Study of Pipeline Gas Production from Coal," *Chem. Technol.* 1, 238 (1971).

(11) *The Economy, Energy, and the Environment*, Joint Economic Committee of the Congress of the United States, Washington, D.C. (September 1, 1970).

(12) H. Lowery, Ed., *The Chemistry of Coal Utilization*, Suppl. Vol., Ch. 21 (John Wiley, New York, 1968).

(13) "Underground Coal Gasification." *Combustion 20* (1949) No. 9, p. 46.

(14) D. I. Mendeleev, "The Future Power on the Banks of the Donets River." *Severny Vestnik* (St. Petersburg) 1888, No. 12, Sec. 11, p. 26.

(15) A. G. Betts, Brit. Patent No. 21674.

(16) G. Lehmann, "The State of the Underground Gasification of Coal," *Bergbau*, 1 (1950) p. 148/150.

(17) O. Glivicky, Communications on the results of a half-scale underground gasification trial from May 1972, Brezno—VUHU, Most, Czechoslovakia.

(18) A. D. Little, op.cit., pp. 17-69.

(19) C. J. Bise, and R. V. Ramaini, "An Evaluation of Underground Mining Techniques for Western Thick Coal Seams," *AIME Preprint* 75-F-341, Sept. 10–12, 1975.

(20) G. B. Glass, "Applicability of the United States Coal Resources to In Situ Gasification," Presented at the First Annual Underground Coal Gasification Symposium, Laramie, Wyoming, July 28–Aug. 4, 1975.

# Theoretical Aspects
# of Coal Gasification

The material in this chapter was excerpted from PB 241 892; UCID-16155; UCID-17007; UCRL-50026-75-1; UCRL-50026-75-2; UCRL-50026-75-3; UCRL-50026-75-4; UCRL-51217; UCRL-51770; UCRL-51835. For a complete bibliography see p 251.

## GENERAL COMPOSITION AND REACTIONS

### Fossil Fuel Thermal Efficiency

The function of practically all fossil fuels in energy production is based upon their combustion in air and the extraction of the heat from the combustion products. Classically, the heating value of the fuel is considered in terms of the heat of combustion per unit weight (solid fuels) or per unit volume (gaseous fuels). This method of quantifying the relative heating value of a fossil fuel probably stems from the fact that most fuels are transported some distance from their source, such as a coal mine, to the user, such as an electricity generation plant. In this case, one desires fuels with a maximum potential energy content in the fuel itself. However, when the fuels are to be used at the site of the source, a possibly more meaningful quantification of the relative heating value might be the sensible heat content of the gaseous combustion products of the fuel when burned with air. These gases are the first-stage working fluid in any power generation system based on fossil fuel combustion; for example, in making high-pressure steam or driving a gas turbine.

Table 2.1 lists the heat content of several fuels, including "low"-Btu gas compositions, which have been obtained in underground gasification experiments. While there are large variations in $\Delta H$(combustion) for various fuels, these variations are not necessarily reflected in the sensible heat of

12

the combustion products. Thus, while $CH_4$ would normally be considered to have a higher heating system than $H_2$ or CO, its combustion products (in air) have less extractable heat content. The surprising result of this method of looking at the energy of fuels is that the "low"-Btu fuel gases, which have heats of combustion about one-tenth that of methane, have ΔH(product) values about two-thirds that of methane. In other words, the "low"-Btu gases may be poor from a fuel transportation view point but good from an on site utilization viewpoint.

Table 2.1 indicates that coal represents the fuel with the highest value of ΔH(product), about twice that of the "low"-Btu gases. Thus, in terms of an in situ gasification process with on site conversion of the combustion energy to electricity, it could be more efficient to completely burn the coal than to attempt the production of a combustible gas. It should be pointed out, however, that the actual thermodynamic advantage of one in situ process over another would depend on many additional factors; for example [1] whether or not the sensible heat of the "low"-Btu gas produced can also be recovered, and [2] the overall efficiency of each process in extraction of the coal energy from underground.

## TABLE 2.1: COMPARISON OF HEATING VALUES FOR FUELS

| Fuel | ΔH (combustion) H$_2$O as Gas | | . . . . . . ΔH (products) . . . . . . | |
|------|---------|---------|---------|---------|
| | kcal/mol | Btu/unit | H$_2$O as Gas, Btu/ft$^3$ (STP) | H$_2$O as Liquid, Btu/ft$^3$ (STP) |
| H$_2$ | 58.7 | 291/ft$^3$ (STP) | 102.2 | 119 |
| CH$_4$ | 201.1 | 997/ft$^3$ (STP) | 95.8 | 104.9 |
| CO | 68 | 337/ft$^3$ (STP) | 118.5 | 118.5 |
| A* | 25.3 | 125/ft$^3$ (STP) | 62.5 | 69.7 |
| B** | 17.6 | 87.3/ft$^3$ (STP) | 57.0 | 61.1 |
| Coal | 138 | 13,000/lb | 119.0 | 123.9 |

*Low-Btu fuel gas composition reported in BuMines Hanna, Wyo., UCG experiment (1).
**Low-Btu fuel gas composition reported in BuMines Gorgas, Ala., UCG experiment (2).

Source: PB 241 892

## Chemical Reactions

The chemistry of coal has been extensively studied. Several excellent texts give the details (3). Although it is known that coal is a polymer, the structure of that polymer has recently been reexamined. The conventional

view is that coal is basically constructed of aromatic/hydroaromatic structures (Figure 2.1) but recently it has been proposed that coal is based on an
adamantane, or modified diamond-like structure (5). More recent work
by Sternberg and co-workers suggests that coal has an acid-base structure
similar to that found in asphaltenes (6).

## FIGURE 2.1: POSSIBLE CHEMICAL STRUCTURE FOR COAL

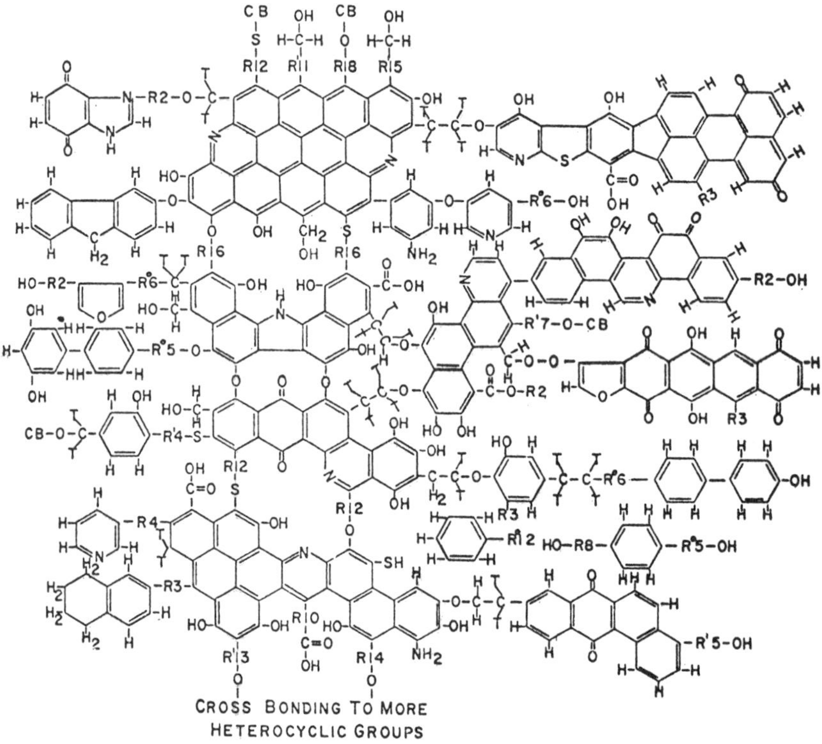

CROSS BONDING TO MORE
HETEROCYCLIC GROUPS

R°N = Alicyclic rings of N carbons.
RN = Alkyl side chain of N carbons.
R'N = Unsaturated alkyl side chain of N carbons.
CB = Cross bonding by O or S to new heterocyclic groups with side chains.
T = Tetrahedral 3 dimensional C—C bonds, C—O bonds and C—S bonds.

Source: U.S. Patent 3,244,615

Because the phenanthrene-like molecule is partially hydrogen-saturated, it is strained into "boat-like" shapes. Nitrogen and nonpyrite sulfur are contained within the rings; oxygen is mostly in the hydroxy form. These elements thus cross-link the basic ring structures into phenol-formalde-hyde-like polymers. Because of the ring strain, the whole structure is so loose that water molecules can occupy space between loosely parallel rings, forming hydrogen bonds with the unsaturated carbon atoms and thus giving additional stability to the whole structure. This water comprises 10-30% of the weight of coal in place and chemically is part of the coal.

Coals are classified into a complex sequence depending on hydrogen-carbon ratio, coking properties, ash content, behavior when heated, and sulfur content. This discussion will consider four broad categories; anthracites, bitumins, subbitumins, and lignites. In this order, the classification is characterized by generally increasing hydrogen-carbon ratios, generally decreasing heats of combustion, and generally increasing oxygen content. Gasification appears applicable to those coals that are in the subbituminous and lignite classifications and that in the "as-received" state have hydrogen-carbon ratios approaching one. This type of coal has a large amount of fixed hydrocarbon and has a more favorable heat balance during in situ gasification.

Since each coal will have a slightly different chemical composition, a generalized set of reactions cannot be written. For the purpose of analysis, the coal in a deposit in the central Powder River Basin of eastern Wyoming, will be discussed.

The formula for the coal indicated in Table 2.2 is $CH_{1.16}O_{0.35}N_{0.015}S_{0.005}$ which can be simplified to $CH_{1.16}O_{0.35}$ for the thermodynamic calculations. This coal has a gram-formula weight of 18.76; this should not be confused with its molecular weight which is 1,000 or more. In the following calculation the "mole" used is assumed to be the simple one shown above.

### TABLE 2.2: ANALYSIS OF THUNDERBIRD COAL

| . . . . Proximate Percent . . . . | | | | . . . . . Ultimate Percent . . . . . | | | | |
|---|---|---|---|---|---|---|---|---|
| Moisture | Volatile | Carbon | Ash | S | H | C | N | O |
| 21.4 | 33.7 | 38.6 | 6.4 | 0.8 | 5.6 | 58.6 | 1.0 | 27.5 |

Source: UCRL-51217

To calculate heat balances and chemical reactions, it is necessary to know the enthalpy $H^0_{298}$ and the entropy $S^0_{298}$ of the coal compound. The enthalpy or heat of formation of coal can be determined from the heat of combustion and the known heats of formations of the reactant oxygen and combustion products. From this calculation, $H^0_{298}$ of this coal is found to be -30.627 kcal/mole.

Estimating the entropy of formation is more difficult and must be done by making assumptions about the chemical structure of coal. By comparison with a number of organic compounds of similar molecular structure, the entropy of the coal can be estimated to be between 5 and 15 cal deg$^{-1}$ mole$^{-1}$. Use of the Latimer rule (12) for estimating entropy, yields a value of 10.24 cal deg$^{-1}$ mole$^{-1}$ for the coal in question. This very important property should be measured experimentally. However, if it is recognized that there may be some shift in the correct operating temperatures if the true value is much different than that assumed, an assumed value can be used to evaluate the heat and equilibrium reactions. Its choice does not affect the heat balance at all. A value of 10.1 cal deg$^{-1}$ mole$^{-1}$ is assumed here.

The reactions of coal can now be examined. Attention is to be directed to four reactions: combustion, pyrolysis, reaction with water, and reaction with hydrogen. Oxygen, water, hydrogen, carbon dioxide, and carbon monoxide contact the coal at various locations. Neither carbon dioxide nor carbon monoxide react with coal, and the reactions are not considered further. The estimated $\Delta H$ for reaction and the free energy, $\Delta F$, are shown in Table 3 for these reactions at temperatures of 500 and 1000°K. Following the usual convention, a negative value of $\Delta H$ means that heat is released and a positive $\Delta H$ means that heat is consumed. Negative values of $\Delta F$ imply that reactions favor the product (right) side of the chemical equation; positive values of $\Delta F$ indicate that the reactant (left) side is favored. Thus, a reaction with positive $\Delta H$ and negative $\Delta F$ proceeds to completion while consuming heat.

The thermodynamic quantities in Table 2.3 show that coal decomposes or reacts with water (reactions 1 and 2) only if heat is supplied at temperatures well above 500°K. Therefore, in this coal and in the absence of oxygen or hydrogen (reactions 3 and 4) the coal would not continue to react and no "runaway" burn could occur. The reaction that produces methane directly from coal (reaction 2) requires that heat be supplied and appears to be somewhat more favored at these temperatures than the pyrolysis (reaction 1). Hence, the direct conversion of coal to methane with water is possible if enough oxygen is supplied to provide the necessary heat via reaction 4. If, on the other hand, reaction 2 is not favored and reaction 1 occurs instead, then methane and carbon would be produced.

## TABLE 2.3: ΔH AND ΔF FOR REACTIONS IMPORTANT TO THE GASIFICATION PROCESS AT TEMPERATURES OF 500 AND 1000°K.

| Reaction | $\Delta H_{500}$ | $\Delta H_{1,000}$ | $\Delta F_{500}$ | $\Delta F_{1,000}$ |
|---|---|---|---|---|
| | . . . . . . . . (kcal mole$^{-1}$). . . . . . . . . | | | |
| (1) $CH_{1.16}O_{0.35} \rightarrow 0.35H_2O(g)$ $+ 0.115CH_4 + 0.885C$ | +9.304 | +11.134 | +2.159 | -0.997 |
| (2) $CH_{1.16}O_{0.35} + 0.535H_2O(g)$ $\rightarrow 0.557CH_4 + 0.443CO_2$ | +10.526 | +12.026 | +2.896 | -3.234 |
| × (3) $CH_{1.16}O_{0.35} + 1.77H_2$ $\rightarrow CH_4 + 0.35H_2O(g)$ | -7.884 | -8.884 | -4.124 | -1.364 |
| (4) $CH_{1.16}O_{0.35} + 1.115O_2$ $\rightarrow CO_2 + 0.58H_2O(g)$ | -98.637 | -97.112 | -103.687 | -107.212 |
| (5) $C + \frac{1}{2}O_2 \rightarrow CO$ | -26.30 | -26.77 | -37.18 | -47.94 |
| (6) $C + 2H_2 \rightarrow CH_4$ | -19.30 | -21.43 | -7.84 | +4.61 |
| (7) $C + H_2O \rightarrow CO + H_2$ | +31.98 | +32.47 | +15.18 | -1.90 |
| (8) $CO + H_2O \rightarrow H_2 + CO_2$ | -9.51 | -8.31 | -4.85 | -0.63 |
| (9) $CO + 3H_2 \rightarrow CH_4 + H_2O$ | -51.28 | -53.87 | -23.02 | -6.51 |
| ×(10) $2CO \rightarrow C + CO_2$ | -41.49 | -40.78 | -20.03 | +1.27 |
| (11) $CO + \frac{1}{2}O_2 \rightarrow CO_2$ | -67.79 | -67.55 | -57.21 | -46.67 |

Source: UCRL-51217

The carbon would then react with water (reaction 5), and a portion of the resulting carbon monoxide would react further with water (reaction 8) to produce enough hydrogen to balance reaction 9, with the same net methane production.

Reactions 3 and 10 would present difficulties if they occurred. Methane production via carbon would be inhibited if reaction 3 occurred. The hydrogen necessary to convert carbon monoxide to methane would be consumed; and the product gas would be a mixture of methane and carbon monoxide. Similarly, if reaction 10 occurred, carbon monoxide would be consumed; therefore, no methane would be produced by reactions 8 and 9. It appears from experiments done in surface methanators that the rates of reactions 8 and 9 are favored under such conditions, whereas reaction 10 is favored under others. It is impossible at present to say which would be most likely in situ. However, if reaction 10 occurred, then when the higher temperature zone reached the product carbon, hydrogen, and carbon monoxide would be produced via reaction 7. These could be combined over the proper catalyst at the surface to form methane. The secondary reactions in Table 2.3 between carbon, methane, carbon dioxide, hydrogen, and water are reproduced from a similar table in Ref. 9.

Several other conclusions can be drawn from this table. No oxygen can survive even at modest temperatures, since reactions 4, 5, and 11 are all exothermic with strongly negative free energies. Reactions 1, 2, and 7 consume energy at high temperature while 6 and 9 or 10 produce energy at lower temperatures. This combination causes the heat to be "spread" through the reaction zone, and helps the combustion zone to "skip" across barren shale zones without the necessity for external attempts at reignition.

All the information in Tables 2.2 and 2.3 can be combined for calculation of an overall energy balance. In the following discussion it is assumed that water and oxygen are added in quantities just sufficient to maintain reaction 2 of Table 2.3 at 700°K. The ambient temperature is assumed to be 300°K and the coal composition is as shown in Table 2.2. Under these conditions, the heat required is that to raise the coal, ash, and water, and to sustain the reaction. The heat supplied must come from combustion of coal. For this calculation it is convenient to express the thermodynamic quantities on a unit weight basis as shown in Table 2.4.

## TABLE 2.4: THERMODYNAMIC INFORMATION
## FOR CALCULATION OF OVERALL ENERGY BALANCE

(1)  1 g coal-in-place = 0.936 coal + 0.064 g ash

(2)  1 g coal + 1.9 g $O_2$ → 2.34 g $CO_2$ + 0.56 g $H_2O$(g)
$\Delta H$ = -5,291 cal/g coal

(3)  1 g coal + 0.513 g $H_2O$(g) → 0.475 g $CH_4$ + 1.038 g $CO_2$
$\Delta H$ = +529 cal/g coal

(4)  1 g coal at 300°K → 1 g coal at 700°K
$\Delta H$ = +255 cal/g

(5)  1 g $H_2O$(l) → 1 g $H_2O$(g) (T = 393°K)
$\Delta H$ = +575 cal/g

Source:  UCRL-51217

Essentially 0.2 gram of coal-in-place must be burned for each gram of coal-in-place converted with water. The products and reactants per ton of coal-in-place are summarized in Table 2.5.

## TABLE 2.5: REACTANTS AND PRODUCTS PER TON
## OF COAL-IN-PLACE

| | - - - - - - - Reactants - - - - - - - | |
|---|---|---|
| Coal-in-place | 1 metric ton | |
| Oxygen | 0.304 ton | |
| Water | 0.306 ton | 306 liters |
| | - - - - - - - Products - - - - - - - - | |
| Methane | 0.370 ton | 19.8 MCF |
| Carbon dioxide | 1.175 tons | 22.9 MCF |
| Ash | 0.064 ton | |
| Nitrogen | 0.0072 ton | 0.22 MCF |
| S* | 0.0058 ton | 0 |

*It is not clear whether sulfur would appear as $SO_2$ or $H_2S$, but $CS_2$ is not found at low temperatures. Either $SO_2$ or $H_2S$ would likely react with the shales and be absorbed before reaching the product line.

Source: UCRL-51217

The calculation above is approximate, since several of the heat capacities have been estimated. The calculation leaves the products at 700°K, but surely some part of the heat will be recovered during the flow of hot gases through the cooler unreacted coal beyond the reaction zone and during the flow of input oxygen and water through the ash and spent material leading up to the reaction zone. The water use has been calculated as if all of the water remained in the coal, but it most likely would vaporize in the hot downstream gases and appear as liquid condensate in the base of the broken zone. In this case it would have to be pumped to the surface and reinjected as needed.

The heating value of the coal used is 9,000 Btu/lb or 19.8 million Btu/metric ton. The heating value of the methane produced is 18.6 million Btu; thus, the heating value of the coal has been largely conserved by this process even though only 46.4% of the carbon is converted to methane. (The small loss is the heat left in hot ashes, shale, and gas.)

The shales between coal beds will use some heat, but under the assumptions that all product heat is supplied from the coal and that products are left at the reaction temperature, this loss does not reduce the reaction efficiency. In a very real sense, the heat balance assumed in arriving at Table 2.5 is a "worst case" from the efficiency point of view.

## THEORETICAL STUDIES

### Convective Instabilities

Theoretical studies (7)(8) of convective instabilities that may occur during vertical in situ coal gasification have been performed. A conjectured advantage of vertical down-flow gasification is that it will create a stable advancing reaction zone. One potential cause of reaction-zone instability is buoyancy-driven convection in the permeable region above the hot reaction surface. Convection cells associated with this type of fluid instability could make the reaction zone burn unevenly and cause control problems during in situ processing.

The reaction zone is modeled as a stationary, constant high-temperature, horizontal plane forming the lower boundary of a porous and permeable layer. The upper boundary consists of a horizontal plane at ambient temperature. Vertical, forced convection of fluid through the layer is assumed. A theoretical hydrodynamic stability analysis is discussed in detail in reference (7), and numerical calculations with applications to coal gasification are presented in reference (8).

The most serious consequence of convective instability is uneven burning at the reaction surface. Convection-cell size is about equal to the thickness of the permeable layer in the absence of forced convective through-flow, but decreases rapidly with increasing flow rate. For typical operating flowrates, convection cells will be of the order of the particle size within the permeable layer and are not likely to produce significant perturbations on the reaction surface. Figure 2.2 shows a machine plot of calculated

**FIGURE 2.2: MACHINE PLOT OF CALCULATED STREAMLINES**

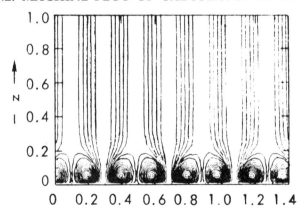

Source: UCRL-50026-75-1

streamlines in the vertical plane at the onset of instability, as drawn by a computer contour-plotting routine. The plot respresents a vertical cross section of the permeable region above the hot burning surface showing multi-cellular convection patterns with a forced downflow convection rate of $10^{-4}$ m/s. The vertical coordinate, z, extends from the burning surface to the top of the coal seam (0 to 1 in dimensionless scale units). The horizontal coordinate, x, has the same scale units. A low, forced convection velocity has been used to illustrate the shape of the convection cell patterns. The cells will be much smaller at expected operating flowrates.

## Factors Influencing Coal Plasticity

Coal Type: The type of coal is the most important factor in coal plasticity. Plasticity is generally minimal to negligible for coals of the subbituminous rank or lower, and for coals of the anthractite rank (9). Coals of the bituminous rank exhibit a range of plastic behavior. Quantitatively, plasticity is evident when the volatile matter on a dry, ash free basis (daf) is lower than 35-40%. It becomes a maximum when the volatile matter is 25-28%. When the volatile matter decreases to 15-18%, plasticity vanishes (10).

Mackowsky and Wolff (11)(12) measured swelling after coking individual particles of a variety of coals. Their results agree qualitatively with the previous statement by Loison (10), that plasticity is a maximum for coal with a daf volatile matter of 25-28%. Dulhunty and Harrison (13) demonstrated the changes in swelling with coal rank.

Increased Heating Rate: Increasing the heating rate increases the plasticity. Swelling and fluidity are seen to increase dramatically with increasing heating rates. The softening temperature changes little, but the resolidification temperature increases, increasing the plastic temperature range.

Dulhunty and Harrison (13) heated particle beds of the different types of coal linearly from room temperature to 700°C in from 4 min to 144 hr. The bituminous coals show increased swelling with more rapid heating. Mackowsky and Wolff (11)(12) measured the changes in particle size after coking individual particles at 600°C at various rates of heating. Swelling was found to increase with increasing heating rates.

Increased Particle Size: Increased particle size increases swelling as demonstrated by Mackowsky and Wolff (11)(12). Various sized individual particles of a variety of coals were coked at 2°C/min in a dilatometer. Particles of 0.2 mm diam or less experience little net swelling when coked individually. However, when they are heated in a bed, as in a dilatometer, they experience considerable swelling due to the fusing together of the particles.

Pressure Effects: The effect of pressure on coal plasticity has not been extensively investigated. The evidence suggests that plasticity does increase with increasing pressure. However, a pressure change of many atmospheres is necessary for a significant effect.

Oxidizing-Reducing Atmospheres: Oxidizing atmospheres decrease plasticity. Conversely, reducing atmospheres increase plasticity. It is well known that weathering in air degenerates the coking properties of coal. Intense oxidation dramatically affects plastic properties (10).

**Plasticity Considerations**

The gasification of plastic coals requires special process design considerations and the use of an oxidation pretreatment step. The surface gasification of plastic coals is always preceeded by oxidation pretreatment to destroy the agglomerating tendencies (14)(15)(16). Even though the agglomerating tendencies are destroyed, the coals can still swell as much as 50%. In surface gasification, swelling does not deteriorate the process.

In in situ gasification, the necessary preoxidation process can be accomplished by using a reverse or countercurrent burn. In a reverse burn, the gases are flowing opposite the direction of flame-front travel, subjecting the preheated coal to an oxidizing atmosphere. In a forward burn, the gases are flowing in the direction of flame-front travel, subjecting the preheated coal to a reducing atmosphere. Gasification could be accomplished entirely by reverse burning, but forward burning is generally preferred since it generates a higher quality gas and permits greater resource recovery.

In addition to preoxidizing with a reverse burn, the process design should avoid the necessity of flow through small cracks and fractures since these would likely be closed by the swelling of plastic coal. Large flow passages can be created by extensive reverse burning, and allowing the roof to collapse into open spaces to form a rubblized bed. Explosives could be utilized to insure roof collapse.

Flow passages in plastic coals can be closed by overburden pressure as well as by swelling. Gasification processes are planned at depths of 300 m (1,000 ft) and greater. A zone of coal that shows any plasticity at all will not be able to maintain open flow passages if it must support the 3 MPa (400 psi) overburden pressure at that depth. Hence, the combustion heating should be localized. This is fortunately the case with most in situ gasification processes. The processes are usually initiated in localized channels. The low thermal conductivity of the coal maintains a localized heating effect. Significant swelling will occur in subbituminous coals during in situ gasification since the heating rates are in the range of 0.1°C/min.

## FUNDAMENTAL PROCESSES AND PROPERTIES

### Reactivity of Chars

Underground coal gasification is in fact underground char gasification. The rate at which gasification can be carried out efficiently depends upon the reactivity of the char, particularly the reactivity with respect to steam. The reactivity of chars has been measured in steam and carbon dioxide. The chars were prepared from Felix No. 2 coal and Roland seam coal.

The chars from both of these coals appear more reactive than any other coal char measured. This is due for the most part to a very large surface area (as large as 1,000 m²/g measured by nitrogen absorption). Both chars are 4 to 5 times more reactive in steam than in carbon dioxide. The fraction of char consumed in flowing steam and carbon dioxide at 1 atm depends upon time (t) and a rate constant (R) according to the following approximate equation:

$$\text{Fraction of ash-free char reacted} \cong 1 - e^{-Rt}$$

The rate as a function of temperature is given in Figures 2.3 and 2.4. The lower line in both figures is for char that was preheated at 800°C before reaction with steam or carbon dioxide. The other data were obtained by making the char at the temperature the reaction rate was measured. Pretreatment at high temperatures decreases reactivity due in part to a decrease in surface area. The change in rates with temperature suggests an activation energy of 50 ± 5 kcal.

### Kinetics of Coal Pyrolysis

It is well known (17)(18) that by measuring product composition as a function of temperature at constant heating rate one can obtain information on the rate of pyrolysis, and the associated activation energy and preexponential constant. Although this is not kinetic data in a rigorous sense, it is sufficient to allow accurate modeling.

Apparatus to study coal pyrolysis under different gaseous environments and heating rates was built (see Figure 8.17 p 189). By using a heated sampling system with a flow by-pass, it is possible to sample all products (gas and tar) without interrupting flow in the reactor. Since the flow of a

FIGURE 2.4: SUBBITUMINOUS CHARS
REACTION RATES IN CARBON DIOXIDE

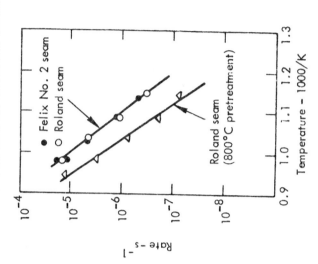

The lower curve shows the reduction in reactivity due to pretreatment at 800°C.

Source: UCRL-50026-75-4

FIGURE 2.3: SUBBITUMINOUS CHARS
REACTION RATES IN STEAM

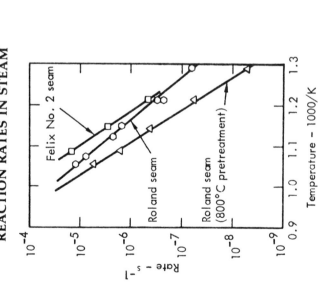

These are the fastest rates ever found for coal chars. The reactivity is decreased by pretreatment at 800°C.

Source: UCRL-50026-75-4

carrier gas through the system is maintained constant (±1%) throughout the experiment, a quantitative measure of gaseous product composition relative to the constant background of carrier gas is obtained. Further- more, the carrier gas can be changed to simulate the pyrolysis environ- ment of an in situ gasifier.

Figures 2.5 to 2.7 show the measured pyrolysis product composition ver- sus temperature for Wyodak coal. A heating rate of ~3.3°C/min and argon carrier gas were used. The volumes of the product gas are normalized to that of the constant-flow carrier gas. Although the greatest volume (moles) of gas is given off at high temperatures ($>500°C$), the largest weight loss occurs below 500°C. This is due to the high molecular weight of the carbon dioxide and $C_n$ ($n>3$) products.

### FIGURE 2.5: EFFECT OF TEMPERATURE ON THE EVOLUTION OF HYDROGEN

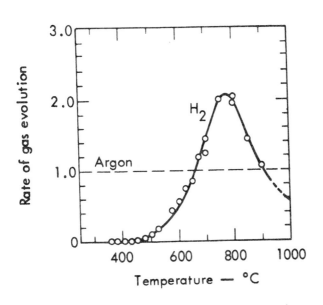

The heating rate is 3.3°C/min, the carrier gas is ar- gon, and the particle size is 1.65 to 3.35 mm (1 unit on the vertical axis is 0.417 $cm^3$/g·min). The rate of gas evolution is normalized to a constant flow of carrier gas.

Source: UCRL-50026-75-3

**FIGURE 2.6: EFFECT OF TEMPERATURE ON THE EVOLUTION OF METHANE, CARBON DIOXIDE, AND CARBON MONOXIDE**

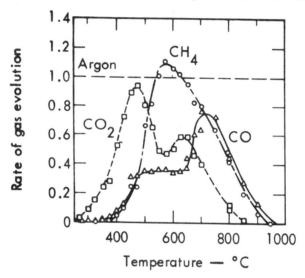

The conditions are as in Figure 2.5.

Source: UCRL-50026-75-3

**FIGURE 2.7: EFFECT OF TEMPERATURE ON THE EVOLUTION OF ETHANE, PROPANE, AND ETHYLENE**

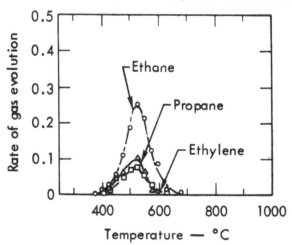

The conditions are as in Figure 2.5, except that 1 unit on the vertical axis represents 0.459 cm³/g·min.

Source: UCRL-50026-75-3

Figure 2.8 is a plot of weight loss vs temperature in an inert nitrogen atmosphere. A comparison with Figures 2.5-2.7 shows the effect of the high-molecular weight products at lower temperatures. A number of authors (17)(19) have used weight loss rather than gas volume and composition as a means of estimating pryolysis kinetics. This can lead to erroneous conclusions and unreasonable values for kinetic parameters.

### FIGURE 2.8: WEIGHT LOSS vs TEMPERATURE

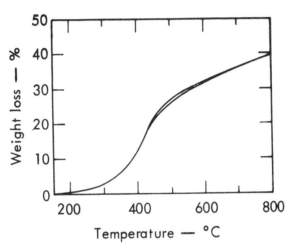

Two separate runs are shown to illustrate reproducibility. The heating rate was 3°C/min and the nitrogen flow rate was 85 cm³/min.

Source: UCRL-50026-75-3

The total quantities of the different gases given off (i.e., the integrals of Figures 2.5 to 2.7) can be measured. From these data one can then predict the average gas composition of gas evolved as a result of pyrolysis during in situ gasification.

Figure 2.6 shows twin evolution peaks for carbon dioxide. The peak at higher temperature is thought to result from the decomposition of the carbonate in the ash, since other results are in agreement with the approximate temperature at which the maximum of this curve occurs.

### Isothermal vs Nonisothermal Kinetic Methods

The best known method of studying reaction kinetics uses isothermal techniques. The change in concentration of a particular reactant or product species is monitored as a function of time at a given constant temperature. By running several experiments at different temperatures, one obtains both the energy of activation and the kinetic frequency factor for the reaction. The mechanism for the reaction is then usually obtained by varying the inital concentration of the reactants and/or in some cases, an intermediate species.

In the case of reactions that occur at high temperature it is often difficult to insure isothermal conditions during the entire course of the reaction. This is because there is an initial heat-up period. For this reason nonisothermal techniques are often used. A detailed description of this technique is given in the literature (20)(21).

### Liquids (Condensibles) from Coal Pyrolysis

In the temperature region of 300 to 500°C liquid products are generated from subbituminous coal. Here the term liquid is used to refer to the total condensible product at 0°C, i.e. water and tar. The water/tar ratio is roughly 3/1 for this coal.

Figure 2.9 is a plot of the weight of liquid material produced as a function of temperature. The tar is a complicated mixture of many different organic compounds as indicated in a gas-chromatography simulated distillation curve. The temperature of the chromatography column can be related to the actual boiling point for a particular n-alkane giving one a boiling point distribution curve for the tar.

### FIGURE 2.9: TAR PRODUCTION AS A FUNCTION OF TEMPERATURE

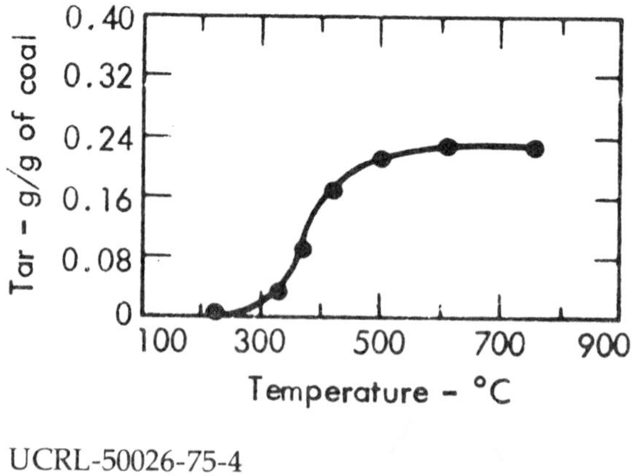

Source: UCRL-50026-75-4

**Solid Products**

The solid residue formed during pyrolysis is a heterogeneous mixture of hydrocarbons (char) and inorganic impurities (ash). The weight loss vs temperature for Wyodak coal was found to be ~45 to 50% at temperatures between 900 and 1,000°C. A distinct change in the chemical composition of the material accompanies this weight loss. This is illustrated in Figure 2.10 where the percentage of carbon and hydrogen in Wyodak coal (or char) is plotted vs temperature.

**FIGURE 2.10: PERCENTAGE OF CARBON AND HYDROGEN IN WYODAK COAL OR CHAR AS A FUNCTION OF TEMPERATURE**

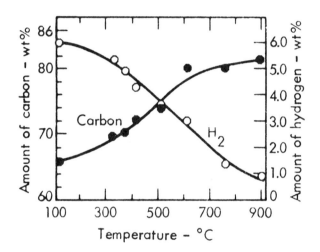

Source: UCRL-50026-75-4

Although both carbon and hydrogen are evolved during the pyrolysis process, much more hydrogen is given off than carbon. At lower temperatures this is due to the large amounts of methane, alkanes, and hydrocarbon tars that are evolved. Since in all of the products the hydrogen/carbon ratio is greater than one, then correspondingly the ratio of hydrogen to carbon must decrease in the char. At higher temperatures (i.e, T > 700°C) large amounts of hydrogen are driven off as the char begins to form crystallites of graphite. This leads to a further decrease in the char hydrogen/carbon ratio. In the limit of complete graphitization the ratio goes to zero.

As a result of the loss of organic material during pyrolysis the ash concentration begins to increase. This is shown in Figure 2.11 for Wyodak coal. At approximately 615°C the carbonates in the ash decompose giving carbon dioxide. This is evidenced by both a reduction in acid-evolved carbon dioxide and a second peak that is observed in the carbon dioxide gas evolution curve.

**FIGURE 2.11: PERCENTAGE OF ASH IN WYODAK COAL OR CHAR AS A FUNCTION OF TEMPERATURE**

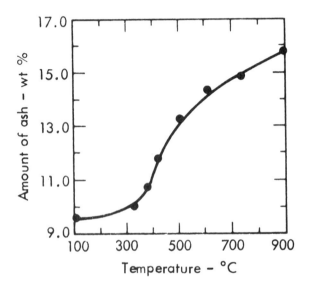

Source: UCRL-50026-75-4

An important physical property of char (in relation to its chemical reactivity) is its surface area. There is little information on surface areas for Wyodak subbituminous coal thus far. Taylor and Bowen (22) recently investigated the reactivity of Wyodak coal char with water and carbon dioxide. They interpret the initial large increase in rate with burn-up in terms of the corresponding large increase in char surface area.

The relationship of the surface area of Wyodak coal char to temperature of char formation has been studied. It is interesting to note that the surface area increases about twenty fold between 300 and 700°C and then decreases. This large increase is surface area near 600 to 700°C has also been reported for chars from pure polymeric materials (23). Many other properties also show maxima in that region. It is unclear whether the evolution of pyrolysis products opens up more surface area or whether the char sites become more activated by release of certain compounds. Attempts are being made to relate these increases to changes observed in other properties.

## REFERENCES

(1) L. A. Schrider, and J. Pasini III. "Underground Gasification of Coal—Pilot Test, Hanna, Wyoming." Presented at 5th Synthetic Pipeline Gas Symposium, AGA, October 1973, 18 pp.

(2) J. L. Elder, M. H. Fies, H. G. Graham, R. C. Montogmery, L. D. Schmidt, and E. T. Wilkins. *The Second Underground Gasification Experiment at Gorgas, Ala. Bu Mines RI 4808*, 1951, 72 pp.

(3) H. Lowery, Ed., *The Chemistry of Coal Utilization*, Suppl. Vol., Ch. 21 (John Wiley, New York, 1968).

(4) G. R. Hill and L. B. Lyon, *Ind. Eng. Chem.* 54, 36-39 (1962).

(5) S. K. Chakrabartty and H. O. Kretschmer, *Fuel* 53 (No. 2), 132-135 (1974).

(6) H. W. Sternberg and C. L. Delle Donne, *Fuel* 53, 172 (1974).

(7) G. M. Homsy and A. E. Sherwood, *Convective Instabilities in Porous Media With Through-Flow*, Lawrence Livermore Laboratory, Livermore, Rept. UCRL-76529 (1975).

(8) A. E. Sherwood and G. M. Homsy, *Convective Instability During In Situ Coal Gasification*, Lawrence Livermore Laboratory, Livermore, Rept. UCRL-51791 (1975).

(9) W. R. K. Wu and W. H. Fredric, *Coal Composition, Coal Plasticity, and Coke Strength*, U.S. Dept. of the Interior, Bureau of Mines, Bulletin 661 (1971).

(10) R. Loison, A. Peytary, A. F. Bower, and R. Grillot, "The Plastic Properties of Coal," *Chemistry of Coal Utilization*, Supplementary Vol., H. H. Lowry, ed. (John Wiley & Sons, Inc., New York, 1963) pp. 150-201.

(11) M. Mackowsky and E. Wolff, "Microscopic Investigations of Pore Formation during Coking," *Coal Science*, Advances in Chemistry Series No. 55, R. F. Gould, Ed. (American Chemical Society, Washington, D.C., 1966).

(12) M. Mackowsky and W. Wolff, "Coking Properties of Coals of Different Grades with Special Regard to Bulk Density and Rate of Heating of Charge," *Erdöl und Kohle-Erdgas-Petrochemie*, 18 (8), 621-625 (August 1965).

(13) J. A. Dulhunty and B. L. Harrison, "Some Relations of Rank and Rate of Heating to Carbonization Properties of Coal," *Fuel 32* (4), 441-450. (October 1953).

(14) A. J. Forney, R. F. Kenny, S. J. Gasior, and J. H. Field, "Destruction of Caking Properties of Coal by Pretreatment in a Fluidized Bed," *I&EC Product Research and Development, 3* (1), 48-53 (March 1964).

(15) S. J. Gasior, A. J. Forney, and J. H. Field, "Destruction of the Caking Quality of Bituminous Coal in a Fixed Bed," *I&EC Product Research and Development, 3* (1), 43-47 (March 1964).

(16) S. J. Gasior, A. J. Forney, and J. H. Field, "Decaking of Coal in Free Fall," *Fuel Gasification,* Advances in Chemistry Series No. 69, R. F. Gould, Ed. (American Chemical Society, Washington, D.C. 1967).

(17) (See for example) H. A. G. Chermin and D. W. Van Krevelen, *Fuel 36,* 85 (1957), and references cited therein.

(18) H. Juntgen and K. H. Van Heek, *Fuel 47,* 103 (1968), and references cited therein.

(19) D. R. Stephens, Lawrence Livermore Laboratory, Internal Document CDTN-72 (1975).

(20) K. H. Van Heek, H. Juntgen, and W. Peters, *Ber. Bunsenges. Physik. Chem. 71,* 113 (1967).

(21) K. H. Van Heek and H. Juntgen, *Ber. Bunsenges. Physik. Chem. 72,* 1223 (1968).

(22) R. W. Taylor and D. Bowen, *Rate of Reaction of Steam and Carbon Dioxide with Chars Produced from Subbituminous Coal,* Lawrence Livermore Laboratory, Rept. UCRL-52002 (1976).

(23) H. Marsh and W. F. K. Wynne-Jones, *Carbon 1,* 269 (1964).

# General Methods
# of Underground Gasification

The material in this chapter is excerpted from PB 209 274. For a complete bibliography see p 251.

The methods for carrying out underground coal gasification can be classified as shaft methods, shaftless methods, and combinations of these two. Shaft methods require some kind of underground development such as shafts, entries, galleries, slopes, or inclines where men have to work underground at some stage in order to prepare the coal seams for subsequent gasification. Shaftless methods do not require much underground work, since preparation is done from the surface invariably involving the use of drill holes to reach the coal seams. Combination methods involve some underground work, but use surface drill holes for the air inlets and gas outlets. Combination methods have been involved in much of the experimental and test work reported in the literature.

The following table is a summary of some of the experimental and commercial gasification installations reported in the literature. It is evident from the entries that an underground gasification system employing shaftless methods can be completely developed and operated with the current state of the technology without having men exposed to the hazards of underground work. However, if for some reason the results of underground gasification become much more attractive by having some limited access to the installations through shafts or inclines, facilities such as drilled shafts can be provided such that only a small risk of exposure of men to dangerous or hazardous conditions would be involved.

## SHAFT METHODS

Extensive experimental work has been done on three shaft methods: chamber, borehole producer, and stream.

33

## Underground Gasification Methods

| | Country | Location | Type | Coal Seam. Thickness (in) | Depth (ft) | Dip (°) | CV Btu/lb | Technique | Type Operation | Linkage | Pattern | Blast | CV Product Gas Btu/scf | Remarks |
|---|---|---|---|---|---|---|---|---|---|---|---|---|---|---|
| A Shaft (underground development) | Russia | Lisichansk | Bituminous | 30 | 100 | 40 | 11,600 | Streaming | Commercial | Gallery | Panels | 30% O₂ | 86 | 6% heat loss |
| | Russia | Gorlovka | Bituminous | 72 | 200 | 75 | 10,000 | Streaming | Commercial | Gallery | Panels | 30% O₂ | 130 | high gas loss |
| | Poland | – | – | 39 | – | – | – | Producer | Experimental | Drill holes | Parallel | Air | 80 | |
| | U.S. | Gorgas | Bituminous | 35 | 25 | 2 | 14,000 | Streaming | Experimental | Gallery | U-shape | Air | 47 | high gas loss |
| | Belgium | Bois-la-Dame | Semianth. | 36 | 525 | 87 | 14,000 | Streaming | Experimental | Gallery | 37° Panel | Air | 56 | 35-45% thermal efficiency |
| B Shaftless (boreholes) | U.S. | Gorgas | Bituminous | 37 | 180 | 2 | 15,000 | Percolation | Experimental | Electro | Square | Air / Oxygen / Water gas | 93 / 195 / 279 | 30-50% gas loss |
| | U.S. | Gorgas | Bituminous | 37 | 180 | 2 | 15,000 | Percolation | Experimental | Hydraulic | X-Pattern | Air | 84 | 44% thermal efficiency |
| | U.S. | Gorgas | Bituminous | 37 | 180 | 2 | 15,000 | Percolation | Experimental | Hydraulic | Straight line | Oxygen | 124 | |
| | U.S. | Gorgas | Bituminous | 37 | 180 | 2 | 12,000 | Percolation | Experimental | Hydraulic | 50' circle | Air | 90 | 40% gas loss |
| | U.K. | Newman-Spinney | Bituminous | 36 | 75 | 8 | 12,800 | Streaming | Experimental | Boreholes | Straight line | Air | 85 | |
| | U.K. | Newman-Spinney | Bituminous | 36 | 100 | 8 | 12,800 | Percolation | Experimental | Pneumatic | Rectangular | Oxygen | 100 | |
| | Russia | Yushno Abinsk | Bituminous | 276 | | 75 | 12,000 | Percolation | Commercial | Pneumatic | Inclined | Air | 130 | 65% heat recovery |
| | Russia | Lisichansk | Bituminous | 30 | 100 | 40 | 11,600 | Percolation | Commercial | Pneumatic | Inclined | Air | 100 | high pressure used |
| | Russia | Moscow Field | Lignite | 72 | 65 | 0 | 4,900 | Percolation | Commercial | Pneumatic | 75' square | Air | 85 | |
| | Russia | Shatsky | Lignite | 120 | 150 | 0 | 4,900 | Percolation | Commercial | Pneumatic | 25 meter grid | Air | – | |
| | Russia | Stalinsk* | Bituminous | 98 | 1,500 | 80 | 14,000 | Streaming | Commercial | Pneumatic | Inclined | Air | – | |
| | Russia | Tula | Lignite | 390 | 180 | 0 | 4,900 | Percolation | Commercial | Pneumatic | 25 meter grid | Air | 105 | |
| C Combination (shafts plus boreholes) | U.S. | Gorgas | Bituminous | 42 | 125 | 2 | 14,000 | Streaming | Experimental | Gallery | Straight line | Air | 70-90 | 4-40% gas loss |
| | U.K. | Newman-Spinney | Bituminous | 36 | 240 | 8 | 12,800 | Streaming | Commercial | Boreholes | Parallel unit | Air | 57 | 84% coal recovery |
| | Morocco | Djerada | – | – | 160 | 90 | – | Streaming | Experimental | Gallery | Panels | Air | – | combustible gas produced |
| | Russia | Lisichansk | Bituminous | 30 | 100 | 40 | 11,600 | Streaming | Commercial | Boreholes | Panels | 43% O₂ | 100 | 47% efficiency |

*Proposed.

## Chamber Method

The chamber, or warehouse, method requires the preparation of underground galleries and the isolation of coal panels (Figure 3.1). The blast of air for gasification is applied from the gallery at the previously ignited face of one side of the panel and the product gas is removed by the gallery at the opposite side of the panel.

### FIGURE 3.1: THE CHAMBER METHOD

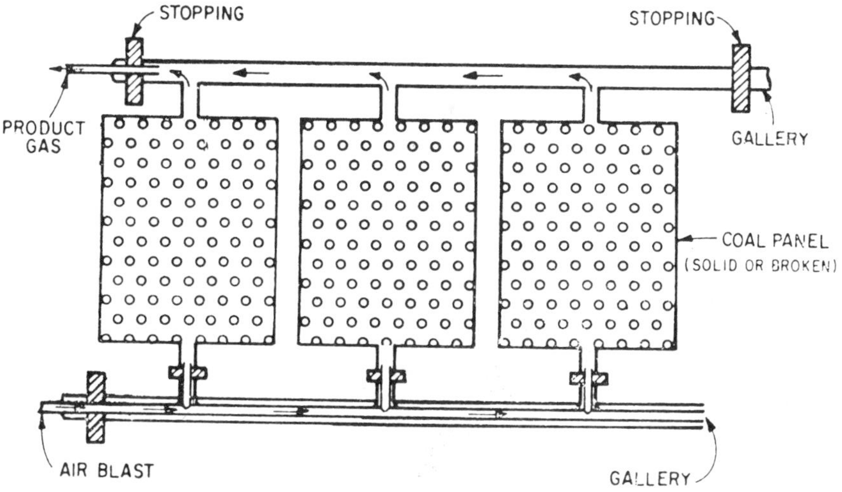

Source: PB 209 274

In one of the first tests in the Moscow area of the chamber method (1), an attempt was made to gasify the virgin coal bed, using its natural porosity and the presence of natural fissures. The rates of gasification were low, and the product gas was of variable composition, sometimes containing unconsumed oxygen. In another attempt a 60 x 22 foot panel containing 1,000 tons of coal was isolated in a 15 foot thick seam, and ignited. Gasification with air proceeded smoothly for 30 days, during which time 200 tons of coal were burned. The product gas had a calorific value of 250-490 Btu per cubic foot, but it appeared that much of the coal was left underground as coke.

A variation of the chamber method in which the coal panel was prefractured by dynamite blasting in holes drilled into the panel was also tried in the U.S.S.R. Subsequent gasification produced only poor gas of irregular quality flowing intermittently, presumably because the coal fractured irregularly, permitting the gasifying air to bypass the reaction zone. Another attempt in the Moscow region (1) involved actually breaking the coal in a panel by hand, followed by gasification. Although this test produced gas of 121-193 Btu/ft$^3$ in heating value, it eliminated only the loading and hauling of coal to outside the mine. It required all other conventional underground development and coal breaking operations to be done. The main result of the work with the chamber method was that it could produce a combustible gas if there were proper coal-gas contact.

## Borehole Producer Method

The borehole producer method also requires underground work, namely, the development of parallel underground galleries. These galleries can be located about 500 feet apart within the coal bed. From such galleries, 4 inch diameter boreholes are drilled about 15 feet apart, from one gallery to the opposite. Boreholes are fitted with valves, and provision is made for electric ignition and for operation by remote control. (See Figure 3.2.)

The gasification is started by igniting the horizontal boreholes farthest from the general access gallery with the control valves closed on all the other boreholes. Gasification air comes down the central inlet shaft or vertical borehole, through the boreholes being gasified, and out the offtake galleries and shafts or vertical boreholes. As one set of horizontal boreholes is gasified the valves on the next set are opened and the gasification proceeds, in this way retreating toward the general access gallery. Careful control of the operations makes possible better utilization of coal and the production of a higher-heating-value gas of a somewhat constant quality. Calorific values of gases produced in a number of experiments (1) using this technique ranged from 160 to 246 Btu per cubic foot.

Work reported by the Bureau of Mines at Gorgas (2) indicates the maximum width of channel that seems practical to produce from a set of boreholes. The maximum attainable width would determine the number of drill holes needed. The borehole producer method was employed except that linking was by hydraulic fracture rather than by drill holes. Figure 3.3 shows a sketch of the probable finished configuration of the cavity (i.e., the situation after gasification of a 3 foot seam). The gasification procedure was the following:

    (a) Two 4 inch diameter steel cased boreholes about 200 feet apart and 185 feet and 163 feet deep were prepared.

## FIGURE 3.2:  THE BOREHOLE PRODUCER METHOD

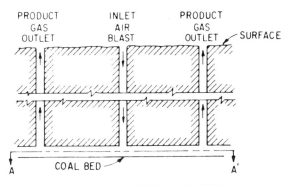

SECTION FROM SURFACE TO COAL BED

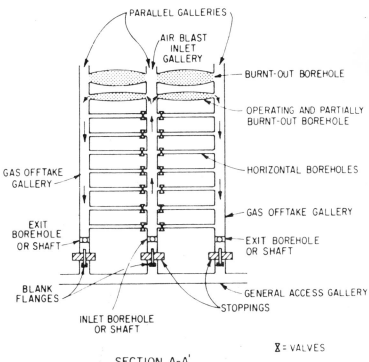

SECTION A-A'

Source: PB 209 274

(b) The coal seam between them was hydraulically fractured with a water gel-sand mixture at 900-1,100 psi and 200-600 gpm for 79 minutes.

(c) The boreholes were linked by backward burning for 50 days (intermediate boreholes were used) with about 150 cfm of air. Ignition was with hot charcoal.

(d) Gasification was carried out for 94 days with about 500 cfm of air.

The probable cross-section size calculation, shown in Figure 3.3, was based on the assumption that the full 3 foot depth of the coal bed was gasified and carbonized along the 200 foot path. The variation in the compositions and heating values of the product gases during the test is shown in Figure 3.4. Calculations indicated that 266 tons of coal were gasified and 370 tons were carbonized, resulting in a 44% thermal efficiency or coal recovery. The drop-off in heating value as the cavity enlarges is quite evident.

Another experiment in Great Britain (3), carried out in a 2.5 foot thick coal bed, produced a reaction channel about 26 feet wide and 400 feet long during one portion of a test in which the coal seam was prepared by the borehole producer technique. The inlet and outlet gas boreholes were 12" diameter cased holes and the 400 foot long initial linking hole was 14" in diameter. The inital air velocity was about 700 cfm, which was increased over the 17 week test period to a maximum of about 3,000 cfm.

## FIGURE 3.3: PROBABLE CROSS-SECTION OF FINISHED CHANNEL AFTER LINKING AND GASIFICATION

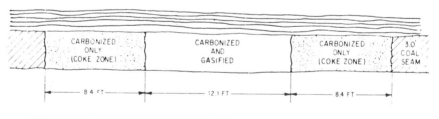

BETWEEN TWO 4-INCH BOREHOLES ±200 FEET APART

Source: PB 209 274

**FIGURE 3.4: VARIATION IN COMPOSITION AND
HEATING VALUE OF PRODUCT GASES
BOREHOLE PRODUCER METHOD**

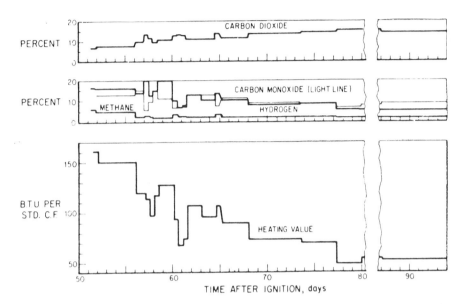

Source: PB 209 274

**Stream Method**

In the stream method the gas/coal contact is perpendicular. A substantial amount of underground development work is required. A general arrangement of a typical installation is shown in Figure 3.5. The method is particularly applicable to steeply dipping coal seams. Inclined galleries are constructed parallel to the coal seam, which are connected at the bottom by a horizontal fire drift. At Gorlovka, in the U.S.S.R., where this method was used with some success, the inclined galleries in the coal bed were 60 yards long and 100 yards apart, so that the panel contained about 12,000 tons of coal. Gasification is started with a fire in the horizontal drift and proceeds up the slope of the coal seam, with air coming down one inclined gallery and the gas going up the other inclined gallery.

One obvious advantage of the stream method is that ash and roof collapse material drops down, fills voided space, and at the same time does not tend to choke off the combustion zone at the burning coal face.

At the Gorlovka site, the stream method was applied to a 6 foot seam,

## FIGURE 3.5: THE STREAM METHOD

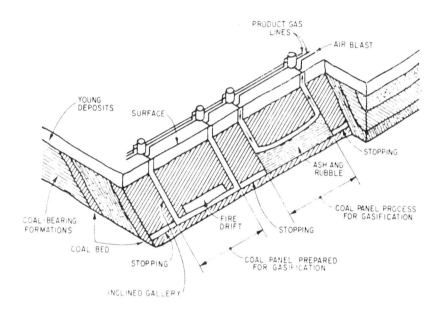

Source: PB 209 274

dipping at 70-75°, and the effect of using oxygen enrichment of the blast was investigated. The researchers reported that the most satisfactory operation was with 27-30% oxygen in the blast. With this amount of oxygen, a gas of 112-146 Btu/ft³ was produced. Another result of the preceding work was the production of a hydrogen-rich gas, when the air blast was interrupted. By running alternately with and without air blasts for periods of 4 to 6 hours, a gas containing about 45% hydrogen with a heating value of 180 Btu/ft³ was produced during the no-blast period. When the air blast was enriched to a 35% oxygen content, the heating value of the product gas rose to 235 Btu/ft³ during the no-blast period, while during the blast period the gas produced had a heating value of 137 Btu/ft³. The use of steam in the blast gas was also reported and a mixture of enriched air and steam gave uninterrupted production of a gas of 295 Btu/ft³ heating value.

A variation of the stream method was used in the experimental test carried out at Bois-la-Dame, Belgium (1). Figure 3.6 illustrates this installation. The coal was a semi-anthracite containing about 13% volatile matter. The coal bed was approximately 3 feet thick, dipping at an angle of 87°, containing approximately 20% ash. In the experimental installation, the fire drift was constructed at an angle of 37°. The objective was to burn the coal panel in a generally sidewise direction instead of upward as in the Russian work at Gorlovka. A combustible gas was produced at Bois-la-Dame on several occasions. But gasification results were poor, and the attempt to burn the coal panel sidewise has been considered to be unsuccessful.

FIGURE 3.6: PANEL No. 1 AT BOIS-LA-DAME

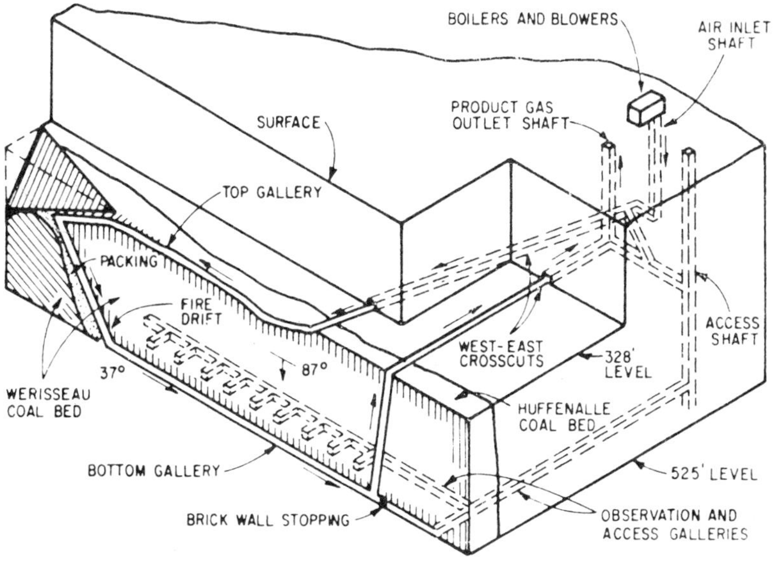

Source: PB 209 274

## SHAFTLESS METHOD

In shaftless methods, all development and gasification is carried out through boreholes drilled from the surface into the coal seam. No underground labor is required. A great variety of borehole spacings and locational patterns have been used, as have been numerous different pregasification and gasification procedures.

The general approach used in the development of shaftless methods has been to make the coal bed more permeable to gas flow between inlet and outlet boreholes. First a linking technique has been employed, then the coal seam has been ignited by passing air, or other gasifying agents, through the inlet boreholes to the outlet borehole.

### Percolation (Filtration) Method

The simplest and the most direct approach to accomplish shaftless gasification of a coal seam is by the so-called percolation, or filtration, method. An installation using this approach is shown in Figure 3.7. The coal seam is penetrated by two or more boreholes some distance apart and gasification takes place between different pairs of holes to the extent required to produce a product gas continuously. For some coals, the method can be made to work using only the natural permeability of the coal bed and the presence of fissures. Some lignites, for example, have a naturally high permeability. This approach, however, has not often been possible with high-rank coals. Thus, in such cases, it becomes necessary to connect boreholes by some linking technique that will increase the permeability and/or fracture the coal seam so that an increased rate of acceptance of gas flow can be obtained.

**FIGURE 3.7: PERCOLATION METHOD**

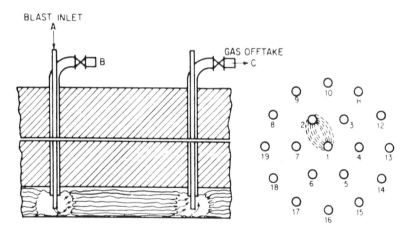

(a) SECTION THROUGH BOREHOLES          (b) PLAN OF BOREHOLES

Source: PB 209 274

As shown in Figure 3.7, the original development and operation of the percolation method involved drilling a number of boreholes 20 to 40 yards apart in a pattern of concentric rings. The method of operation was to fire the coal at the bottom of a borehole, electrically or with glowing charcoal, and to maintain combustion in the area around the base of the hole by supplying air or oxygen through the central pipe of the borehole. Initially, the product gases passed up the annulus and were withdrawn at the surface. However, as the coal was heated, fissures formed and it became possible to pass the gases through the coal seam to an adjacent borehole. One percolation installation that operated in Russia in the Moscow region is a good example of a shaftless gasification system. The plant operated on coal of the Novo-Basovsk bed, which is a brown coal, containing approximately 30% moisture, 37% ash, and having a heating value of 4,900 Btu/lb. The bed ranged in thickness from 1½ to 15 feet, but coal thicknesses less than 3 feet were not gasified.

The natural permeability of the coal bed is used, although it may be increased pneumatically by use of high pressure air, or by electrolinking carbonization. When the pneumatic method is used, backward burning is applied in order to develop the reaction zone. By 1955–56, gas production at the plant had reached approximately 90% of designed capacity. The annual gas production was reported to have been 15.6 billion cubic feet of gas, a quantity sufficient to operate a 15,000 kw electric generating station, if one assumes that 10,000 Btu is required to generate 1 kwh of electric energy.

The plant used a number of boreholes spaced approximately 75 feet apart, arranged on a square pattern. A production of 45 million cubic feet of gas per day, i.e., at the annual production rate noted above, required at least four generations, each composed of not less than eight boreholes operating simultaneously. Each generator is prepared for regular operation in two stages, i.e., first four boreholes in a row are ignited and linked to each other, forming a 225 foot fire front; next, four boreholes are ignited in a second row, 75 feet distant, and linked to the corresponding borehole in the fire front of the first row (see chapter on International Developments, Soviet Efforts). The result is a generator composed of four parallel gasification passages, each 75 feet long, terminating at right angles in a fire front 225 feet long. Gasification proceeds by blowing air to low pressure. The air enters the boreholes in the second row and the gas is removed from the boreholes in the first row.

While gasification is being conducted, linkages are being established between a third row of boreholes and the second row. When the gasification and establishment of this linkage are completed, gasification begins between the third and second rows. Additional rows of boreholes are thus successively linked and gasified over the life of the installation. Thus, for a 1,000 MW power plant, the gasification area might be about 3 miles wide and the gasification direction would advance at right angles over a total distance representing coal adequate for a 20 to 30 year life of the generation equipment. The actual distance would depend on coal recovery and seam thickness. The capital requirements would be low.

There are really no other basic shaftless methods since the drilling of boreholes is the basis of shaftless access to seams. However, there are numerous variations of the percolation method which involve various borehole sizes, locational patterns, linking, and gasifying procedures.

### Variations in Percolation Method

A good example arises from the work of the British Ministry of Fuel and Power beginning in 1949. The first installation (3) involved a shaftless technique somewhat different than the percolation method. Figure 3.8 shows a sketch of the work at Chesterfield in Derbyshire, where tests were conducted in flat-lying coal beds. A four-inch-diameter hole was drilled from an exposed coal face through the inclined coal bed to link the vertical boreholes drilled from the surface. The installation was operated using a reversing flow of air and gas through the vertical boreholes and during part of the trials a combustible gas having a heating value of 60 to 100 Btu/ft$^3$ was produced.

### FIGURE 3.8: FIRST GASIFICATION SYSTEM AT CHESTERFIELD

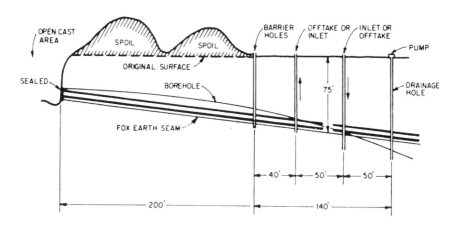

Source:  PB 209 274

Another shaftless method, having possible future interest for selected seams, involves the use of nuclear explosions to fracture coal beds before gasification. The concept is to fracture a 200 foot long series of coal beds in Wyoming at a depth of 1,000 feet. After fracturing, the coal would be ignited and burned under controlled conditions, with oxygen or air-oxygen mixtures provided through one borehole and with gas extraction through the borehole originally drilled to place the explosive. Resources of deep coal amenable to this type of development are tremendous. One hundred square miles are said to contain 20 billion tons. This gasification approach is probably limited to very deep coal seams where there is a substantial thickness of coal. A trial has been proposed, but has not yet been carried out.

## COMBINATION METHODS

Methods involving the combination of shafts and boreholes have been studied and tried in a number of cases. One reason is the obvious advantage of easy direct inspection of the result of experimental work.

### Underground Gallery and Surface Drill Holes

The first set of trials in the United States at Gorgas, Alabama, involved a U-shaped underground gallery and surface drill holes (Figure 3.9). The installation for the second set of trials was similar. It used straight-line underground galleries connected to the surface by a series of boreholes.

The Gorgas work was done in a 34 to 46 inch coal seam of high-volatile A bituminous rank. In the first test in the U-shaped channel, runs were made with air which produced a product gas of 46.8 Btu/ft$^3$ average heating value. When using oxygen-enriched air a heating value as high as 134.5 Btu/ft$^3$ was obtained. The work at Gorgas employed the stream method for flat-lying coal seams. Combustible gases were produced at Gorgas at certain times, but in general, there was excessive bypassing of the reaction face by the air blast, subsequent mixing of unreacted air with combustible gases with combustion of this mixture prior to leaving the installation.

### Underground Openings and Boreholes

Another combination system that used underground openings and boreholes from the surface was the French work at Djerada in Morocco (1). The installation is shown in Figure 3.10. The intention was to ignite the coal panel at the bottom and, in much the same manner as in the Russian work at Gorlovka, to burn the panel from the bottom upward. As shown,

## FIGURE 3.9: PLAN VIEW OF A GASIFICATION MINE FIRST GORGAS EXPERIMENT

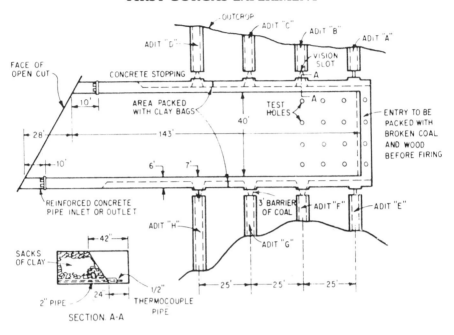

Source: PB 209 274

## FIGURE 3.10: STREAM METHOD INSTALLATION AT DJERADA

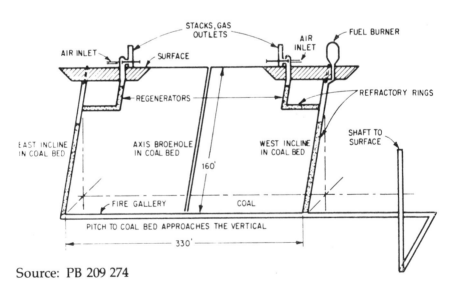

Source: PB 209 274

the fire gallery is connected to the surface by inlet and outlet shafts, the shafts in turn are filled with refractory rings and the passages are fitted with regenerators so that in using reversing flow operation some of the sensible heat in the product can be saved and used for preheating and inlet air. The operation in this experiment was continuous for 5 months. A combustible gas was produced, the oxygen content was nil and remained so throughout the experiment, but the percentage of combustible components varied widely. Operational difficulties, such as leakage at the surface and high resistance to flow within the passages, were encountered.

### Shaft, Gallery and Boreholes

An interesting combination method was also tried in the British experiments at Newman Spinney (near Chesterfield) in the final trial of the program, No. P.5 (3)(4)(5). The general layout of the underground development for the work, and the plan of the installation, is shown in Figure 3.11. It consisted of driving a gallery 225 feet long from a shaft; then drilling four parallel 14 inch-diameter boreholes, 75 feet apart and 400 feet long from the gallery rising with the seam. These boreholes were intersected at their dead ends with vertical holes drilled from the surface. The system was ignited in the gallery at the mouth of each borehole, where air entered through three vertical 12 inch-diameter holes.

### FIGURE 3.11: VIEW OF THE GALLERY
### TRIAL No. P.5

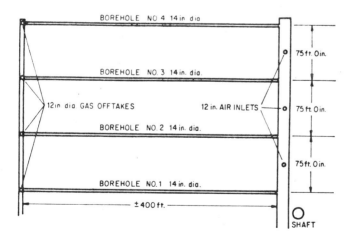

(continued)

**FIGURE 3.11: (continued)**

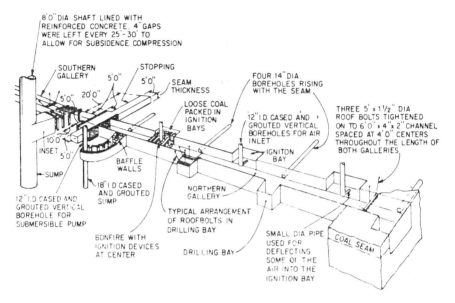

Source: PB 209 274

The system operated for 118 days, gasifying about 900 tons of coal and producing gas with an average calorific value of 57 Btu/ft³ from a 2.5 foot-thick coal bed of 12,500 Btu/lb heating value coal. Air rates of up to 13,000 cfm were used for the total air supply through the three vertical holes. The operating results are summarized in Figure 3.12 in terms of the variation with time of the main operating conditions during the trial, namely: thermal yield, heating value, maximum air pressure, and average air rate. Figure 3.13 shows the estimations of the areas of coal consumed after 32 hours, 58 hours, 89 hours, and after about 2,800 hours when the test was finished.

After completion of the 118 day test, the researchers estimated that 84% of the coal had been exhausted, or gasified. The 16% of coal ungasified was largely accounted for as pillars remaining between gasification boreholes at the gas outlet. The researchers believed that extractions of 90% could be achieved in a commercial operation. There was some evidence that a few pillars had been bypassed by the gasification reactions, and it was thought that a revised spacing of boreholes might prevent this. The work also indicated that at high gas velocities some coke was dislodged from the fire face and brought to the surface. This coke had to be removed in cyclones before the gas could be burned.

**FIGURE 3.12: PERFORMANCE CURVES—TRIAL No. P.5**

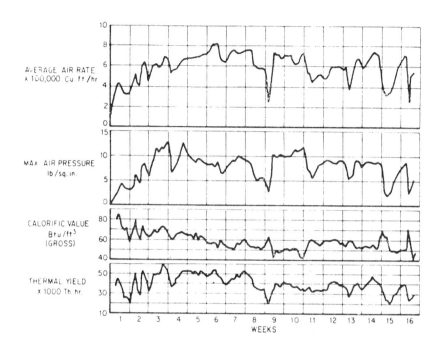

Source: PB 209 274

**FIGURE 3.13: GROWTH OF VOID DURING GASIFICATION
TRIAL No. P.5**

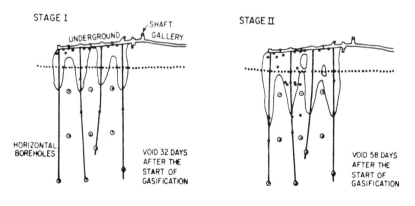

(continued)

**FIGURE 3.13: (continued)**

STAGE III

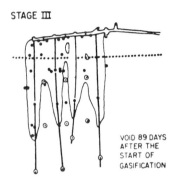

VOID 89 DAYS
AFTER THE
START OF
GASIFICATION

STAGE IV

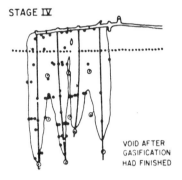

VOID AFTER
GASIFICATION
HAD FINISHED

VERTICAL BOREHOLES
⊙ ORIGINAL HOLES FOR GASIFICATION
● PROBE HOLES

Source PB 209 274

## REFERENCES

(1) J. L. Elder, "The Underground Gasification of Coal" pp. 1023-40 (Chapter 21) in *Chemistry of Coal Utilization,* Supplementary Volume, John Wiley and Sons, Inc., New York, N.Y. 1963.
(2) J. P. Capp, K. D. Plants, M. H. Fies, C. D. Pears, and L. L. Hirst, *Underground Gasification of Coal: Second Experiment in Preparing a Path Through a Coal Bed by Hydraulic Fracturing.* Bur. Mines Rep. Invest. 5808 (1961).
(3) A. Gibb et al., *Underground Gasification of Coal.* Pitman and Sons, Ltd., London, 1964, 205 pp.
(4) J. S. Baxter, "Underground Gasification Trials," *Colliery Eng.* 39 (466), 511-6 (December 1962).
(5) F. E. Warner, J. Szekely, and R. S. H. Mah "Underground Gasification of Coal," *Chem. Eng.* No. 163, A68-78, (October 1962).

# U.S. Programs

The material in this chapter is excerpted from CONF 75-1171-1, TID-3349-05345, PB 256 155 and TID-24825. For a complete bibliography, see p 251.

## GENERAL DESCRIPTIONS

The four in situ methods most extensively described are:
1. LVW—Linked Vertical Well—linkages (flow paths) between boreholes
2. LWG—Longwall Generator—directional drilling
3. TPB—Thick Packed Bed—explosive fracturing of coal underground followed by gasification
4. SDB—Steeply Dipping Bed

The general consensus, is that the LVW and TPB procedures are applicable to the thick western coal seams and that the LWG and SDB procedures are applicable to both western and eastern coal resources.

In general, eastern seams are thinner and it is only rarely that a seam over 10 feet thick occurs. The underground gasification of the average horizontal eastern seams will therefore be costlier when compared to the thick western seams, if one uses vertical holes. It is for this reason that the LWG concept using only a few directionally drilled holes appears to be the promising technique for eastern coals or for thinner seams anywhere. Some eastern coal seams, particularly the Pittsburgh seam, contain significant amounts of methane that can be liberated and recovered prior to an in situ gasification operation and some western seams may contain recoverable methane.

51

## Research Projects

The federal government first sponsored an underground coal gasification program at Gorgas, Alabama, conducted by the Bureau of Mines. It reinitiated underground coal gasification research in 1971–72 by funding a U.S. Energy Research and Development Administration (ERDA) study at the Morgantown Energy Research Center (MERC) which led to the LWG concept involving directional drilling of parallel holes through the coal bed. In 1972, projects were also initiated at the Laramie Energy Research Center (LERC) Laramie, Wyoming and at the Lawrence Livermore Laboratory (LLL) Livermore, California.

Two National Science Foundation-RANN Division sponsored projects are being conducted by universities. The University of Texas at Austin project is designed to evaluate the Texas Lignite beds which lie below the strip mining depths. In addition to NSF, the Texas Utilities Services Co., Shell Oil, Continental Oil, Mobil Oil and State of Texas are supporting this project. The University of Alabama underground coal gasification project is designed to evaluate some 2-3 foot thick bituminous coal beds within a 50 mile radius of Tuscaloosa, Alabama. In addition to the principal NSF-RANN Division funds, the state of Alabama and the Alabama Power Company are supporting this project. Other university programs at West Virginia University, Penn State University, University of Wyoming, University of New Mexico, and University of Kentucky are investigating various phases of underground coal gasification.

Certain phases of underground coal gasification are being investigated by other national laboratories including: carbonization at Oak Ridge National Laboratory; char reaction kinetics at Argonne National Laboratory; pyrolysis at Hollofield National Laboratory; development of underground coal gasification instrumentation for subsurface monitoring with some laboratory testing and a field monitoring project at Hanna, Wyoming by Sandia Corporation, Albuquerque. A U.S. Bureau of Mines project at Bruceton, Pennsylvania is focused on the direct combustion of coal beds to utilize the sensible heat of the gaseous products of combustion.

Increasing interest in underground coal gasification from the industrial sector is most encouraging. The AMAX Corporation is interested in testing the Longwall Generator Concept and possibly other techniques in the midwestern Illinois coals on a joint ERDA-AMAX basis. A very significant step has been taken by Texas Utilities Services, Inc. On March 16, 1975 the Texas Utilities entered into a contract with V. O. Licensingtory of the Soviet Union to transfer the Russian technology to them for application to the lignite beds of East Texas. Other industrial groups are forming to study the underground coal gasification process, such as the one coordinated by Resource Science Co. and including American Oil, DuPont, Pacific Gas and Electric Co., Rocky Mountain Energy Co. and TRW.

The major portion of the TRW proposed program involves development of various concepts for the high-temperature gasification of coal underground and production of electrical power from low Btu gas generated underground or high Btu pipeline quality gas produced by processing medium Btu gas generated underground. The program encompasses early field experiments, pilot-scale operations, and large-scale field demonstrations of the complete process from coal bed to consumer product; scientific, engineering, and economic analyses and study efforts that precede and parallel field work; and the laboratory and bench-scale efforts leading to and supporting other portions of the program. Also considered are some advanced concepts and techniques other than the major gasification techniques to extract energy from coal. Examples are the self-advancing gasifier, coal liquefaction, and mine pillar recovery. The program provides for continuation and expansion of existing projects, possible initiation of development of an additional gasification concept, increase in supporting technical effort, more attention to advanced ideas, and greater emphasis of scale-up problems.

The details of the major ERDA programs are presented here. Since the Steeply Dipping Bed method has no proponent yet, it is described under Alternative Methods in the next chapter.

## LERC GASIFICATION PROJECT (LVW)

On the basis of field demonstration, the Laramie Energy Research Center Project is the most advanced in the U.S. today. As early as 1972, gas volumes ranging from 50,000 to 3,000,000 standard ft³/day (SCF/D) were produced with heating values ranging from 30 to 465 Btu/SCF. Over at least a 4 month period, an average volume rate of 1.6 million SCF/D with average heating value of 130 Btu/SCF was produced. In the first Hanna experiment the vertical borehole method was used with first a forward and then a reverse burn in the 30 foot thick Hanna coalbed. Energy recovery efficiency was estimated between 30 and 50%.

The second Hanna experiment was for studying coalbed permeabilities and reverse combustion linking using the vertical well method. Over a 38 day period an average of 1.9 million SCF/D injection of air yielded 2.7 million SCF/D of product gas with an average heating value of 152 Btu/SCF. The Sandia Corporation in conjunction with LERC, as is described later, has used elaborate instrumentation techniques to monitor the second Hanna experiment. Details of the project follow.

## Location

Wyoming is the largest coal-bearing region in the United States. Estimates of remaining coal resources in Wyoming at depths less than 3,000 feet are in excess of 136 billion short tons. At depths of 0 to 6,000 feet this estimate would be raised to one-half trillion tons of coal or 17 percent of the National total. The coal-bearing regions of Wyoming are shown in Figure 4.1 along with the Hanna coalfield, the site of the current underground gasification of coal program. This site is near the town of Hanna, Carbon Country, Wyo. The land to conduct the pilot field test was provided by the Rocky Mountain Energy Company, a subsidiary of the Union Pacific Railroad Company. Factors favoring this location are depth, thickness, and quality of coal; proximity of power, communications, and supplies; availability of land and water; and some previous geological knowledge of stratigraphy and structure provided by the Union Pacific Railroad from five coreholes drilled through the coal in 1929.

### FIGURE 4.1: WYOMING COAL-BEARING AREAS

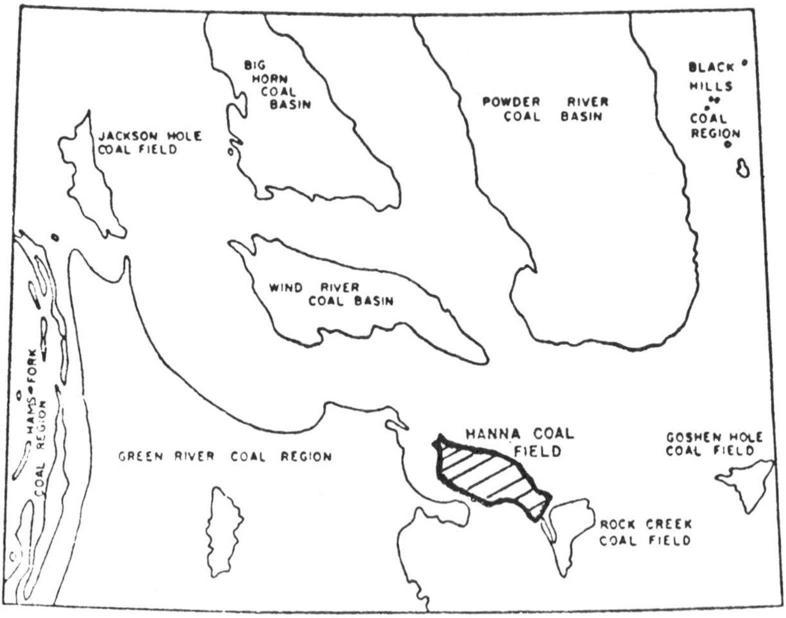

Modified from Berryhill (1950)

Source: TID-3349-05345

## Geology and Stratigraphy

The Hanna Basin covers about 1,000 square miles and contains an accumulation of 35,000 feet of sediments of which up to 28,000 feet may have intermittent coal-bearing strata. The Hanna No. 1 seam found near the base of the Hanna Formation was the coalbed selected for gasification.

The Hanna Basin was subjected to considerable tectonic activity resulting in the present configuration of numerous structural anomalies within the basin. The area of the experiment lies about one eighth mile northwest of the axis of one of the small synclines created by the tectonic movement. The Hanna Formation, containing the Hanna No. 1 coalbed consists of about 7,000 feet of sandstone, siltstone, shale, coal, and thin limestones. As part of the delineation studies prior to gasification, the Federal Bureau of Mines cored part of the first well drilled at the test site. Of the 114 feet cored, the upper 65 feet are interbedded gray siltstone and shale with 15 feet of shale above the coal. The 30 foot thick coalbed was found about 400 feet below the surface with 4 feet of shale and sandstone immediately below. Figure 4.2 is a stratigraphic cross section of the test site showing structure and stratigraphy of the coalbed and its enclosing strata. The experimental site lies on the flank of a comparatively narrow syncline plunging to the northeast.

## FIGURE 4.2: HANNA COAL GASIFICATION SITE WELLS

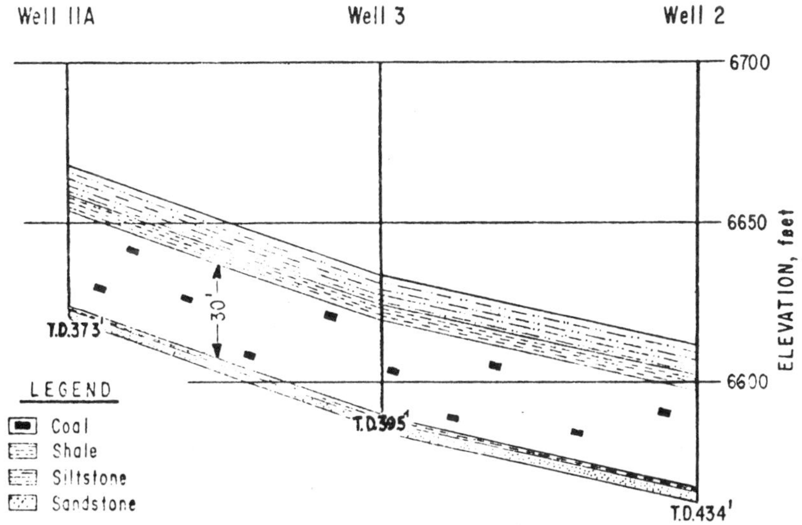

Source: TID-3349-05345

## Oriented Core Studies

In November 1972, the Hanna No. 2 well was drilled and a three inch oriented core was taken from 11 feet above the coal to a point 8 feet below the coal. In total, 16 wells were drilled over an area of about 4 acres. In the spring of 1973, an additional oriented core was taken 1,000 feet northeast of well No. 2. The primary purpose in obtaining the oriented cores was to study directional properties and to determine the dominant direction of gas flow. Knowledge of the preferred flow directions is important to achieve the most advantageous placement of drill holes for liberated gas capture. Since the directional flow properties of the coal are governed by a number of interacting physical properties, several types of measurements were made on the oriented core. These measurements included orientation of coal cleats, directional permeability, directional tensile strength, point load induced failure, and directional sonic velocity tests. The directional permeability and joint trends for the two core holes indicate a preferred northeast-southwest flow direction in the coal.

## Stimulation Treatment

To establish subsurface communication of gases and fluids, hydraulic fracturing of the coal bed using the "sand-frac" process was initiated to increase subsurface permeability. Using well 3, the coal was stimulated with 32,000 gallons of gelled water containing 16,000 pounds of sand at a rate of 1,300 gpm. The formation accepted the material with no apparent "break-down" noted. After fracturing, an impression packer run found four vertical fractures in the wellbore. The major fracture was 1.3 inches wide; three minor fractures were all less than 0.6 inch wide. Air injection tests showed the air acceptance of the coal increased fivefold with the major flow in the southwest direction thus supporting the oriented core studies indicating preferred flow in that direction.

## Ignition

The coal was ignited in well 3 with a propane burner. A forward burning technique in the direction of the major fracture trend was used in an attempt to originate gas production in wells 5 through 8. Additional wells around the ignition well were also drilled to surround the combustion zone. Air injection into well 3 was continued for about 2 months, and product gas was collected from the surrounding wells. Within a few weeks after the initiation of air injection, it became apparent that a large portion of the air (about 90 percent) was not being recovered from any of the producing wells. The majority of the air was therefore bypassing the coal and reappearing through wells which were drilled but not completed with casing and cement to prevent direct communication. The borehole casing in well 3 was probably burned off, permitting communication with the silty sandstone above the coal.

Carbonization tests performed on the Hanna coal at various temperatures resulted in the typical gas composition shown in Figure 4.3.

**FIGURE 4.3: CARBONIZATION REACTIONS PERFORMED ON HANNA COAL No. 1, HANNA, WYO.**

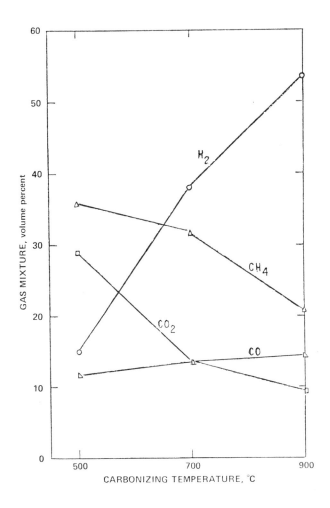

Source: TID-3349-05345

As the temperature is increased, the amount of methane and carbon dioxide decreases. The production of methane is, of course, advantageous since it contains the largest potential heating capacity, approximately 1,000 Btu/SCF; carbon monoxide and hydrogen contain only 300 Btu/SCF. The low temperature reaction (500°C) will not, however, recover a very large portion of the coal potential energy but will result in driving off the volatiles thus leaving about 70 percent of the coal as coke. The following is a summary of the carbonization tests performed for the Hanna coal.

|         | - - - Carbonization Temperature- - - | | |
|---------|--------|--------|--------|
|         | 500°C  | 700°C  | 900°C  |
| Coke, % | 71.9   | 62.1   | 58.3   |
| Gas, %  | 5.1    | 13.9   | 17.3   |
| Tars, % | 23.0   | 24.0   | 24.4   |

With a rise in temperature, more of the latent energy in the coke is utilized and more energy is available as gas. At higher temperatures, the percentage of carbon monoxide and hydrogen should increase in proportion to the reduced amount of carbon dioxide produced. This type of reaction is a good indication of achieving the desired gasification reactions. Although methane will also decrease as carbon monoxide and hydrogen increase, the latter reaction products are still the most desirable for obtaining higher in-place efficiencies. Ideally, it may be possible to achieve a multi-stage process of low temperature devolatilization, eventually followed by high temperature gasification.

## Analysis of Results

Results have been encouraging since flow rates in excess of 2,000,000 SCF/day of 100 to 200 Btu gas have been achieved. Control of this flow, however limited, should be further investigated in order to chart well productivity or performance. Initially, a well produces a high Btu gas from volatilization. This cycle is followed by carbonization and possibly gasification. It has been further shown that switching a well from injection to production can increase flow rates. A good example of this was illustrated when wells 9 and 15 were converted to air intake wells. By utilizing these two wells along the periphery of the gasification site, a backward burn was initiated and drew the combustion zone outward from the center walls. The heating value of the gas and gas volumes have been improved considerably since that time. This combination of gas production and flow conditions needs further investigation to enable predetermination of the life cycle of the wells.

On January 19, 1975, this work was transferred to the U.S. Energy Research and Development Administration. The problems of past tests were avoided and encouraging results were obtained. No gas leakage from the reaction zone was observed. Gas production rate and gas heating value were relatively stable for a 5½ month period. During this period approximately 20 tons of moisture-free coal were gasified per day; energy balance calculations showed 3.5 times more energy produced than consumed, and comparison with an air-blown surface gasifier showed similar energy recovery efficiencies. A second experiment to further define process feasibility, if successful, would lead to design and construction of a 15- to 30-MWe pilot plant, and successful pilot plant operation would lead to design of a commercial demonstration plant by 1980.

## THE LLL COAL GASIFICATION PROCESS (TPB)

### Basic Concept

The ultimate objective of the LLL program is to establish an in situ coal gasification system which is applicable to the processing of deep (500 to 3,000 feet) thick seams and which is sufficiently well defined to enable scaling up to commercially sized projects. This requires defining the preparation (fracturing) and process control schemes in sufficient detail so that economic extrapolations can be made with confidence. The procedure consists of using chemical explosives emplaced through an array of drilled holes to provide initial fracturing and establish necessary permeability in the seam. The permeable bed is then ignited and gasified with an oxygen steam reactant stream at 500-1,000 psi, similar to a packed bed reactor process. The recovered product gases are upgraded in a surface facility to a high Btu content compatible with commercial pipeline quality. A model of the LLL process, based on a coal deposit in the Powder River Basin, Wyoming, is shown in Figure 4.4. A block diagram of the procedure is presented in Figure 4.5.

The details of the many techniques and studies especially designed for and applicable to this process are in other chapters. What is presented here is a basic summary of the process.

### Process Summary

After fracturing, holes are drilled to the bottom of the fractured coal zone and some of the charge-emplacement holes are reentered to gain access to the top of the zone. These wells and injection holes are cased to pre-

## FIGURE 4.4: LLL IN SITU COAL GASIFICATION CONCEPT

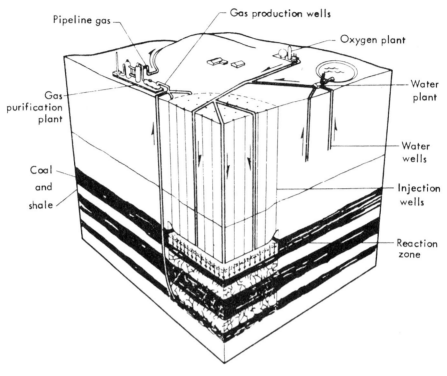

Source: TID-24825

## FIGURE 4.5: BLOCK DIAGRAM OF LLL IN SITU COAL GASIFICATION PROCESS

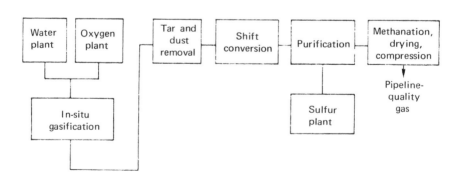

Source: TID-24825

vent unwanted water entry and gas loss. Next, oxygen is injected and combustion is started at the top of the fracture zone. After good combustion has been established, the oxygen is replaced with an oxygen-water or oxygen-steam mixture. The process operates similarly to an underground packed-bed reactor—the shattered coal is gasified with steam and oxygen just as in a conventional high Btu coal gasification plant.

The oxygen is used to burn some of the coal to generate the heat necessary for the endothermic gasification reaction of steam and carbon to produce carbon monoxide and hydrogen. The gases produced underground— primarily methane, carbon monoxide, carbon dioxide, and hydrogen— are treated in a surface facility to produce pipeline quality gas. A permeable, fractured coal bed in an otherwise impermeable medium should permit intimate mixing of coal and reactants, allowing heat transfer and reactant access to the coal. Of course, underground flow patterns must be considered carefully to avoid channeling around the coal, combustion of product gases, or plugging of the permeable reaction zone with water or coal tars—all of which would lead to fluctuating product flow rates, low heating values, and poor resource recovery.

The impermeable surroundings (at depth) should minimize leakage of reactants and product gases from the fractured zone. The 500 to 3,000 foot operating depth for the packed-bed approach is based on process economics. Thick coals at depths less than 500 feet are good candidates for strip mining. Also, methane concentration in the product gas is very sensitive to operating pressures. As depth increases, chemical high explosives become less effective; at some point, other fracturing methods need to be explored. However, for the range of 500 to 3,000 feet, fracturing with an aluminized ammonium nitrate fuel-oil explosive appears to be a good choice. A major objective of the research and development program is to determine the most effective explosive emplacement technique and firing plan to achieve permeability.

Surface facilities for pipeline gas production at the proposed Powder River Basin pilot plant include an oxygen facility, a water plant, and a gas treatment plant. Water will be pumped from subsurface aquifers, stored in a pond, and pumped underground. Unlike surface gasification techniques, which require high quality water, brackish subsurface water should be adequate for the underground portion of this process. In situ gasification should use less high quality surface water, which is already in scarce supply in many coal bearing areas.

The packed-bed process should have little environmental impact. Surface wastes should not be a problem since waste or contaminated water will be used for the underground process. Sulfur-bearing gas can be reduced to elemental sulfur in a Claus plant. Also, because no mining is required, there will be no permanent disfigurement of surface terrain, other than some surface settling as the underground coal is gasified. Experience with subsidence from oil, gas, and water prodcution and from potash and coal mining suggests that surface subsidence will be very gradual and will not have a catastrophic impact on surface plant life. The technical and environmental effects of subsidence will be studied closely, however, as a part of the pilot plant R&D program.

Maintaining continuous 250-MM SCFD ($7.1 \times 10^6$ m³/day) gas production by the packed-bed process will require, annually, preparation and gasification of a coal bed approximately 750 feet in diameter, or a larger number of smaller diameter zones. Projected overall resource recovery cannot be determined yet. This will depend on the maximum dimension possible for gasification and on the width of the coal barrier that must be left between gasified zones to serve as the walls of the underground gasifier. Assuming square fractured regions, the minimum volume of coal available for gasification per unit depth is $a^2/(a+b)^2$, where a is the width of the fractured zone and b is the thickness of coal between zones. In addition to reaction zone size and coal wall thickness, overall resource recovery will depend on the percentage of coal converted in the fractured zone and on the efficiency with which the underground product gas is upgraded to pipeline quality. For a 750 foot diameter reaction zone, geometric resource recovery is 51% for a wall thickness of 300 feet and 78% for a wall thickness of 100 feet.

**Future Research Program**

In general, the laboratory and field programs are comprehensive, and many of the anticipated problems have been considered. Obviously, problems arise during implementation which were not anticipated, but it is to be expected that priorities will be established as the program proceeds; for this reason contingency planning and flexibility are essential. The following evaluation is of general considerations as opposed to details of preconceived problems.

Coal-Fracturing Program: The objective of the coal-fracturing program is to accurately predict the extent and distribution of explosive induced fractures in coal and to correlate these fracture patterns with permeability. A major approximation is the assumption of classical continuum mechanics for the prediction of failure phenomena in coal. Even with this assumption, and assuming that realistic in situ values of the elastic parameters and failure criteria are obtained, the resulting fracture pattern still cannot be related to permeability.

The coal block and outcrop tests are a logical path to take before actual deep tests, but the relevance of the results is in question. The fact that, during one of the coal block tests, the explosive actually migrated into the coal indicates the type of initial fracturing which exists in block tests and outcrop tests and which makes any comparison of observed phenomena with model predictions a very difficult process.

The above evaluation places considerable doubt on the utility of the models for the prediction of permeability, but some effort should still be directed into modeling capability. It is suggested that the fracturing program should include studies of hydraulic or pneumatic fracturing techniques in combination with explosives. There are major problems with all these methods, but, because of the importance of the fracturing program, all avenues should be thoroughly explored.

Coal-Processing Program: The problem of simulating in situ conditions in the laboratory is a very old one in the earth sciences. Packed-bed reactor results are essential for the basic understanding of the combustion mechanism but may not correctly resemble in situ tests. Alternatively, it is very difficult to conduct coal block tests at approximate in situ stress conditions. Within these constraints, the approach taken in the coal-processing program is comprehensive and necessary to the project.

In the operation of reactor experiments or combustion tube tests and in the analysis of the results, the following questions should be considered. Will the heat losses from the system be small enough to obtain satisfactory results? Will axial heat transfer in the tube wall and the surrounding material significantly affect the results? Is the length of the reactor sufficient to obtain or approach pseudo-steady-state conditions? Will this length allow observation of the influence of coal tars which will be deposited on the coal ahead of the reaction zone?

The use of quartz tubes for the reactor vessels will allow visual observations of the reaction process, which will be helpful in the comprehension of the complex phenomena, and these tests will be a useful complement to tests in a heat loss controlled reactor capable of operation at higher pressure.

In the applications of in situ gasification, water as a liquid or as steam will be injected into the coal seam along with the oxygen. The possibility of problems due to oxygen and water separation could become important as in fire-water-flooding operations in oil reservoirs. Both the requirement for water or steam and the possible problem of water-oxygen separation should be investigated, but neither is particularly suited for laboratory experiments.

The coal block experiments will be very valuable for improving the understanding of the process, particularly for separating the effects of the retorting zone and the combustion-gasification zone. The programmed inlet conditions could overcome the problem of insufficient length, but it is not clear how to determine if the programmed inlet conditions are correct.

The materials development program is very important, but it should be noted that corrosion problems encountered in fireflooding operations in oil reservoirs often do not not appear in the laboratory combustion-tube tests. Testing of various materials in production wells during the field tests will be essential to confirm laboratory results.

Field Program: The field program is the most important part of the research, since it is in the field where the utility of in situ exploitation of coal will be proved or disproved. A major unknown factor in in situ gasification is the severity of the problem of subsidence over the retort chamber. Not enough emphasis in the research program has been placed on this very important potential problem. Additional investigations should be performed to evaluate subsidence and how the results of the field tests could be applicable to other sites. A full core from surface to total depth should be obtained so that ground movement predictions may be made and so that future events may be correlated with the overall geology.

The measurement of in situ permeability before and after explosive fracturing is an important part of the field program. However, measurements of communication between wells in a horizontal direction will not necessarily provide information on the desired vertical permeability. In addition, it is even more important to determine the variation in vertical permeability in the horizontal direction. It is not clear at this time how this permeability will be measured. One suggested procedure which will add useful data to the other permeability measurements is to utilize two boreholes packed off at different depths. The results will then be a function of both horizontal and vertical permeability.

For the purpose of evaluating the packed-bed concept, the field program is well conceived and comprehensive. However, because of the formidable problems which will be encountered during field trials, it is important that contingency planning be conducted in case the concept is not successful. An example of such planning is the evaluation of test sites for alternative methods of gasification.

## MERC COAL GASIFICATION PROCESS (LWG)

### Basic Concept

The Morgantown Energy Research Center (MERC) underground coal gasification project was founded in 1971 on the basis of what was thought to be at the time a new way of accessing the coal bed known as the Longwall Generator Concept (Figure 4.6). This scheme was designed to utilize the natural directional properties of the coal bed and allow an enhanced choice of the direction for the gasification front to proceed. As with the longwall mining scheme, greater total resource recovery efficiencies should also be realized. In many parts of the country, the drilling of vertical wells every 50 feet would not be environmentally acceptable or perhaps legally feasible due to complications in obtaining surface land acquisition rights. The use of long directionally drilled holes could greatly alleviate this problem.

The basic Longwall Generator Concept is also perhaps the most versatile scheme for accommodating different modifications for directional flow control, controlled roof collapse and other techniques such as hydraulic fracturing. Similar longwall schemes form the basis for future underground coal gasification plant designs in the Soviet Union. Thus, in many ways MERC was coincidentally in the position of advancing the best technique yet developed and proven in Russia after 30 years of commercialization. No evidence has been found that anyone, including the Russians, has utilized the natural bed directional characteristics in field development plans. The same also applies to petroleum recovery. Perhaps the reason the Russians have not made use of directional bed properties (if they have not) is that most of their coal beds have large dip angles. The directional properties such as permeability become much more important in planning sites with near horizontal beds like a large percentage of those in the United States.

Numerous other schemes such as hydraulic fracturing between straddle packers to form extended vertical fractures perpendicular to the boreholes axis can be useful in the recovery of methane, or other bed preparations ahead of the gasification front.

FIGURE 4.6: IDEALIZED VIEW OF THE REQUIRED PROCESS TO
GASIFY COAL UNDERGROUND FOR POWER GENERATION

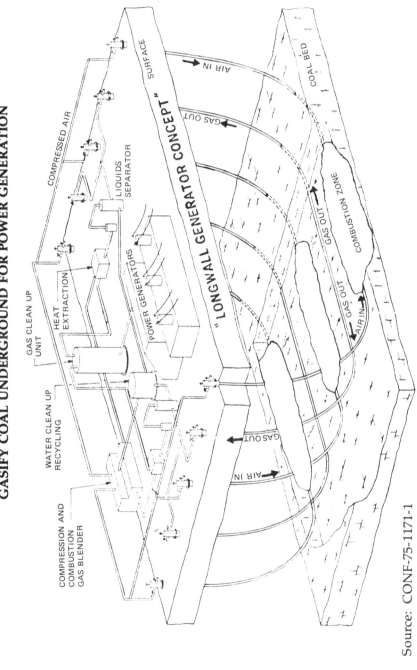

Source:  CONF-75-1171-1

## Alternative Method

Another way of developing the Longwall Generator Concept may be to use conventionally drilled vertical wells and drill small horizontal pilot holes through the coal bed in preferred directions with a high energy laser beam. These small pilot holes up to an inch in diameter would be enlarged by hydraulic fracturing or liquid explosives to provide the desired fracture volumes.

Numerous mechanisms influencing the laser drilling process in coal are being investigated. Some basic computations and modeling of this advanced technology concept are being done at MERC following some successful preliminary coal drilling experiments. Results of the "quick and dirty" drilling feasibility tests indicate a maximum drilling rate of 1 3/8 inch per second in lignite with less than a one kilowatt carbon dioxide CW laser beam. The actual drilling cost was $3.00 per hour. Oil shale, anthracite, bituminous, subbituminous and lignite coals were preliminarily evaluated.

Additional experiments are planned to utilize air, oxygen and nitrogen injection to assess combustion assist-purge and inert gas pruge, continuous and intermittent techniques. Existing laser technology and numerous commercial laser systems have been assessed and found adequate for process evaluation and development.

Work is in progress to develop the laser drilling process in coal and oil shale to optimize mass transfer velocities or drilling rates. Such a capability could provide numerous alternatives to coal resource recovery methods. If proven feasible, it could also have substantial economic advantages over other processes of accessing coal beds for dewatering, methane drainage, and perhaps, safety and communications.

## Research

Other phases of underground coal gasification at MERC include laboratory simulation studies, material property evaluations at elevated temperatures and mathematical modeling. The most abundant coals considered as candidates for underground coal gasification, including bituminous, subbituminous and lignite, are being evaluated under different conditions and schemes in the MERC simulation laboratory.

# Proposed Alternative Methods and Improvements

The material in this chapter is excerpted from PB 209 274, UCRL-50026-75-1, UCRL-51217, UCRL-51676 and various patents as noted. For a complete bibliography, see p 251.

## BLIND BOREHOLE-FILL SYSTEM

Three variations of the blind-borehole technique are possible, depending on the coal seam thickness. The methods and their applicability would be little, if at all, affected by the slope of the seam. Actually steeper slopes could make the stowing easier.

For thinner coal seams, say 2 to 10 feet thick, curved boreholes would be drilled traveling perhaps 1,000 feet along the coal seam (Figure 5.1). The borehole would be fitted with a concentric pipe for its entire length. Gasification of the surrounding coal would occur at the end of each borehole when the oxidizing gas entered through the central pipe. The product gas would leave through the annulus. As gasification proceeded and the voided space increased, filling would be carried out by introducing and suspending an inert fill material in the inlet gas stream, which then would enter the voided space at the velocity of the entering gas. Its momentum would send the material toward the far wall thereby filling the cavity. The expectation is that the filling would reach the roof strata so that it could act as a support. A number of such borehole installations would be installed. As one strip of coal was being gasified, the adjacent stip would be started. The gasification of the entire system then would occur in a staggered fashion.

Before such a method could operate successfully, one would need a hole-casing material that would disintegrate at the temperature of the burning coal face and that would resist abrasion by the fill material. Also filling materials would have to be identified and application techniques developed, that would be capable of roof support adequate to prevent collapse. Most importantly, however, an extension of present technology for drilling curved boreholes would be needed to allow the maximum diameter to be increased and travel distances in coal seams to be increased.

### FIGURE 5.1: BLIND BOREHOLE-FILL SYSTEM
### (FOR THINNER SEAMS)

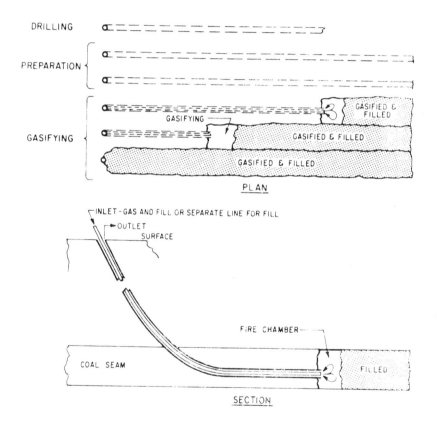

PLAN

SECTION

Source: PB 209 274

For thicker seams, say 10 to 30 feet thick, branching curved holes would be drilled and gasification would take place by a modified streaming method (Figure 5.2). Separate oxidizing-gas inlets and product-gas outlets would be used instead of annular pipes. As in the case for thinner seams, filling would take place as the reactions proceeded with filling material piling up in the back of the cavity. The development problems also are similar.

### FIGURE 5.2: BRANCHED BOREHOLE-FILL SYSTEM
### (FOR THICKER SEAMS)

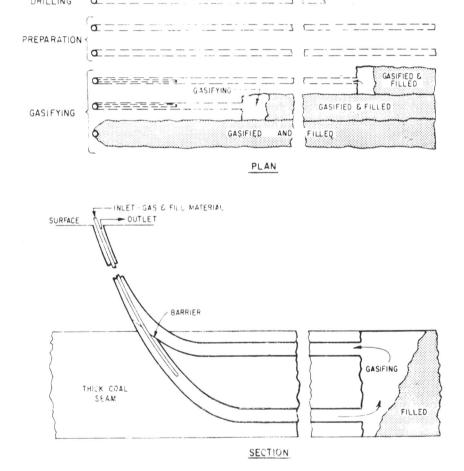

Source: PB 209 274

For very thick seams, say 30 to 200 feet, or for vertical seams, simple easy-to-install vertical boreholes would be drilled and provided with annular pipes. Filling would be aided by gravity (Figure 5.3). Vertical holes might be spaced in plan on a square pattern approximately 50 feet apart.

### FIGURE 5.3: VERTICAL BLIND BOREHOLE SYSTEM (VERY THICK AND VERTICAL SEAMS)

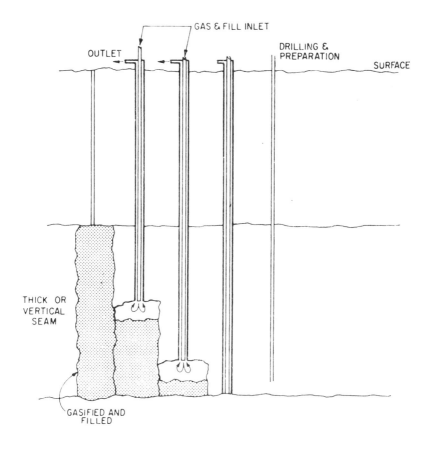

Source: PB 209 274

## SHAFTLESS BOREHOLE-PRODUCER METHOD

Two variations of the borehole-producer method are possible, depending on the method of filling the cavity. Both would be applicable to flat and to mildly dipping seams of thickness up to, say, 10 feet.

### Two-Borehole, Multiple-Unit, Batch-Filled Method

In the two-hole, multiple-unit, batch-filled method a large number of simple two-borehole systems would be installed and operated in a pattern designed to recover most, if not all, the fuel values in the coal bed. The steps would be as follows:

1. Large boreholes (e.g., 14 to 16 inches in diameter) would be directionally drilled from the surface and horizontally for long distances in the seam. A pair of boreholes, perhaps 1,000-1,500 feet apart, would be joined to form the borehole-producer system. The portion of the boreholes within the strata above the coal seam would be cased to prevent gas leakage and water infiltration.

2. The coal between the holes would be ignited and gasified. During gasification, the quantity of oxidizing gas would be continually at an optimum level, that is one that corresponds to the continually increasing size of the cavity. The ultimate size of the cavity was to be determined by the size of the boreholes, the coal seam thickness, the pressure and volume of the oxidizing gas, and the level of heating value in the product gas.

3. When the cavity reached its maximum size, the reaction would be stopped. At this point the cavity should consist of a completely gasified channel, perhaps 30 to 40 feet in overall width, the sides of which are sections of coked-coal, each section perhaps 8 to 10 feet wide. The roof should be intact, since the voided space opened should not be large enough to cause caving.

4. The cavity would now be completely filled through the boreholes with a cement-sand mix of materials having similar properties. The filling technique would still have to be developed experimentally. Drilling of additional small-diameter boreholes from the surface to the cavity might also be required. The filling technique could be a pneumatic method using selected filling materials. A material that expanded as it set would be ideal.

5. After the filling had been completed and the material (if a cementing agent is used) had set, the adjacent panel was to be linked and gasified as above. The sequence would be repeated until the entire coal bed had been gasified and the voided spaces filled. This method would provide for recovering the coke left ungasified from the previous operation in the adjacent panel.

## Multiple-Borehole Continuous-Filling Method

In the multiple-borehole continuous-filling method, essentially similar to that proposed by Sears (1), long horizontal holes would be drilled into the coal seam from the surface by directional drilling techniques (Figure 5.4). A somewhat similar approach was proposed in Russian work for near-surface seams at Kholmogorsk (2).

### FIGURE 5.4: MULTIPLE-BOREHOLE CONTINUOUS-FILLING

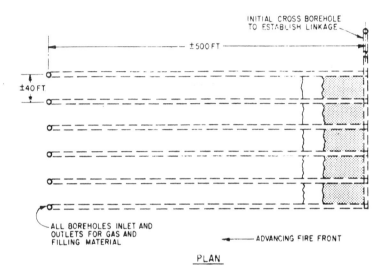

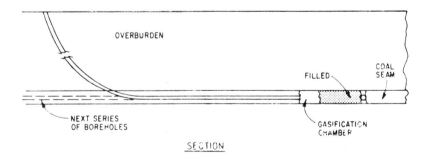

Source: PB 209 274

A series of parallel, directionally drilled and cased, boreholes would be prepared, perhaps 30 to 40 feet apart, and as long as practical (e.g., 500 feet). Positive linkage would be established by drilling one hole direction-ally in a perpendicular direction to the others such that it intersects a number of the parallel holes near their terminals. For example, some 12 parallel holes could be lined this way by one 500 foot-long cross hole. Alternatively, adequate linkage might be established by simultaneously operating the 12 parallel holes as blind boreholes until the individual cavities joined together. Gasification would take place in a continually advancing broad fire front by selectively using the parallel boreholes as gas inlets or outlets, as required, to keep the fire front under control.

To avoid collapse of roof strata, support would be provided as gasification proceeded by continually feeding filling material suspended in the ox-idizing gas as required through the boreholes used as inlets. The amount of filling material suspended would be adjusted to replace the volume of coal burned (less perhaps the volume of ash left behind). The type of filling material best suited to this use would have to be developed. A de-sirable material might be one that tended to fuse and expand at the pre-vailing temperature.

Because of its momentum while traveling in the inlet gas stream, the filling material would have to spray out and away from the fire front and adequately fill the voided space. Proper selection of boreholes as inlets and outlets and control of gas velocities and solids-loading in the gas, should enable the size of the combustion chamber and of gas-coal contact to be controlled, while at the same time preventing roof collapse.

### Shaftless Streaming Method with Stowing

The shaftless streaming method with stowing would be applicable only to sloping seams so that the stowing operations would be facilitated. Ash from the burning coal would fall downward into the voided space as would the stowing materials. The method had been proposed for use in the USSR (2), but probably without the use of stowing for preventing roof collapse and for providing better combustion control.

Figure 5.5 illustrates a possible operation on a near-vertical seam. The di-rectionally drilled holes would permit the vertical seam to be developed and gasified with oxidizing gas and fill material introduced and product gas withdrawn through appropriate holes. The filling would be contin-uous to prevent caving.

## FIGURE 5.5: SHAFTLESS STREAMING METHOD WITH STOWING

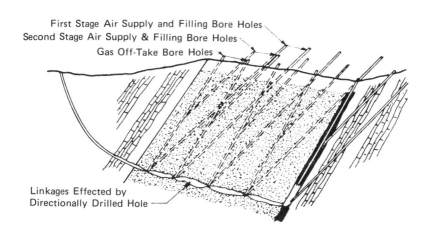

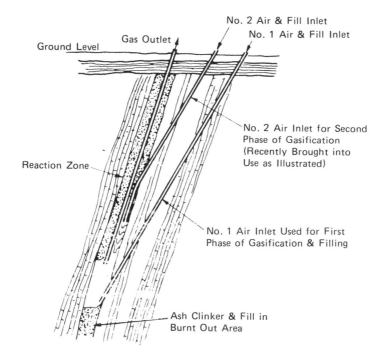

Source: PB 209 274

## STEEPLY DIPPING COAL BEDS

Steeply dipping coal beds are particularly attractive for in situ gasification for two basic reasons: [1] They are generally less economical to mine by conventional techniques than are horizontal seams, and [2] they have the potential for rapid commercialization because of the simplicity of the process, and because of a demonstration of technical feasibility that has taken place in the USSR (See chapter on International Developments). This process is being operated at a semicommercial level in Yuzhno-Abinsk (Siberia) and results from these tests have been published. Product flow rates have averaged 40 million-SCFD of 120 Btu/SCF gas.

The process operates as follows: Injection holes are drilled vertically through the overburden to the deepest region of the coal to be processed (see Figure 9.1 p 223). For thick seams, the injection holes can be drilled "under" the seam, making them insensitive to roof subsidence. Uncased exhaust holes are drilled along the coal seam midway between the injection holes. A manifold is burned out at the bottom to connect the injection and exhaust holes. (The sequence of these operations can be varied as required.) Gasification is then initiated by igniting and injecting either oxygen-steam or air through the injection holes while exhausting the product gas through the exhaust holes. The gasification zone then moves up the rise of the seam, removing essentially all the coal.

The process described has a number of favorable features, some of which are:

1. The hot, dry, and partially pyrolyzed coal immediately above the void zone should easily fall, forming a thick and highly permeable packed-bed of coal.
2. Two reaction zones are formed in series; a packed-bed below the void zone, and the uncased exhaust holes. This gives a redundancy in reaction zones and should result in uniform product gas quality.
3. There should be excellent resource recovery due to the nature of the flow geometry.
4. The system is unlikely to plug as a result of the swelling or plastic flow of coal. In the upper region, the coal is penetrated by large, open, drilled holes, which would be difficult to plug. In the lower region, a packed or rubble bed is formed by coal falling into a void zone. This results in large fraction of void space and makes plugging unlikely.
5. When explosives are used, they are used to produce rubblized

coal that falls down into the void space. Such a physical configuration minimizes the amount of explosive needed for rubblizing the coal.

Although absolute economics are difficult to establish, the simplicity of this approach should result in a process that is economically attractive relative to other alternatives applicable to steeply dipping coal beds.

In the U.S., this type of coal formation is located primarily in Colorado, the Pacific Northwest, and the Appalachian region. Although it represents a small fraction of the total coal resource in the United States, it is still of very significant size.

## CONCEPT FOR IN SITU GASIFICATION OF DEEP COAL DEPOSITS

**Proposed Concept**

The following method appears to represent a possible way by which deeply buried coals can be converted to methane through chemical reactions with oxygen and water. This was the forerunner of the LLL project.

The coal beds would first be shattered in place with explosives. Only the region to be processed would be broken in any one explosion: subsequent blasting would be done to ensure that fractures would not reach a previously processed region. Thus, to achieve high efficiency it would be necessary to drill and blast as large a region as could be processed at one time. In that way only minimal amounts of coal would be left unreacted.

Figure 5.6 shows a portion of a typical blast-hole arrangement. A pattern of 24 inch drill holes spaced about 60 feet apart should break up the coal-containing area. It has been assumed that the explosive used in this case is an ammonium-nitrate/aluminum/diesel-oil mixture. About 600 tons of coal should be broken by each ton of explosive. In suitable coal deposits, several thousand tons of explosives would be emplaced and detonated at one time.

The second step in the process would be to drill access holes to the top and bottom of the shattered region. Some of the explosive loading holes probably could be reused for access and instrumentation if care were taken in stemming them. These access holes would be cased to prevent unwanted water entry.

## FIGURE 5.6: PROPOSED ARRANGEMENT OF EXPLOSIVES

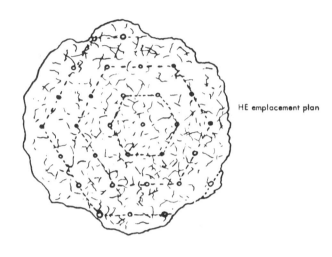

HE emplacement plan

Plan View

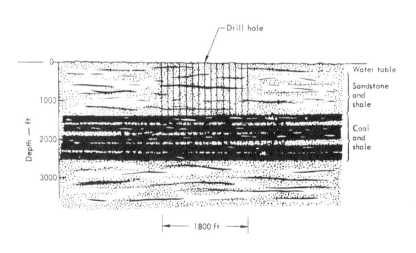

Vertical Section

Source: UCRL-51217

The third stage would involve injecting a small amount of oxygen and starting combustion near the top of the broken zone with a methane flame. When the temperature in the combustion zone reached about 700°K or higher, the oxygen injection would be mostly replaced with water so that the basic reaction would be between coal and water to produce methane and carbon dioxide. To provide heat to sustain methane production, the oxygen/water ratio should be about unity. Back pressure would be built up by restricting the outlet flow until the pressure in the shattered zone was equal to the hydrostatic potential in the coal-bearing zones. In this way little unwanted water entry or gas escape should occur.

Water injection would be continued from the top, and product gases would be withdrawn from the bottom, until all the coal in the shattered zone had reacted. When the hotter zone is at the top, the reaction front is stable and there is no tendency to bypass unreacted coal. This method has been demonstrated with pilot-scale retorts at Laramie (3), and has proved to be more stable than upward or lateral burning. In the region below the reaction zone carbon monoxide is expected to react with water to produce additional methane, carbon dioxide, and heat. Both the reacted and unreacted shale are effective catalysts for the carbon monoxide/water reaction and are also powerful scavengers of sulfur oxides, hydrogen sulfide, and any other acid vapors.

The product gas should be a mixture of methane, carbon dioxide, and water vapor with traces of nitrogen, hydrogen, and carbon monoxide. The removal of water and carbon dioxide, which is a standard and simple procedure, should yield a high quality pipeline gas.

Contrasted with surface-plant processes for coal gasification, this method does not involve mining the coal and does not require a large plant. Heat is conserved because of the low thermal conductivity of rocks and the fact that the inlet gas and water flow through the spent shale and ash, removing their heat. Little or no sulfur, heavy metals, or fly ash should accompany the product because of the effective filtration provided by the long vertical column of broken rock as well as the chemical reactions mentioned above.

Contrasted with other in situ methods for gasification, this method might yield high-quality gas (high heating value), and theoretically, very little coal should be bypassed. The proposed method should have very limited effect on the environment; however, there would likely be eventual surface subsidence.

## Functioning of the Process

In addition to the broken coal, as depicted in Figure 5.6, the major elements of the process are oxygen supply, water supply, gas-purification subsystem, and pumping system. The plant is shown schematically in Figure 5.7. The oxygen plant is assumed to be a standard cryogenic unit, and the water plant probably would consist only of a system for storage and pumping. The gas-purification plant is less well defined because the form in which sulfur could be produced is uncertain. It seems most likely that hydrogen sulfide would be formed in the ground; if that is the case, however, it would be absorbed in the shale by reaction with carbonates and no sulfur gases would be produced. This is a very important and environmentally favorable consequence of in situ processes as contrasted with surface coal gasification methods. Carbon dioxide can be removed by scrubbing or by expansion cooling. There are also several alternative methods.

### FIGURE 5.7: FLOW SCHEMATIC FOR IN SITU COAL GASIFICATION

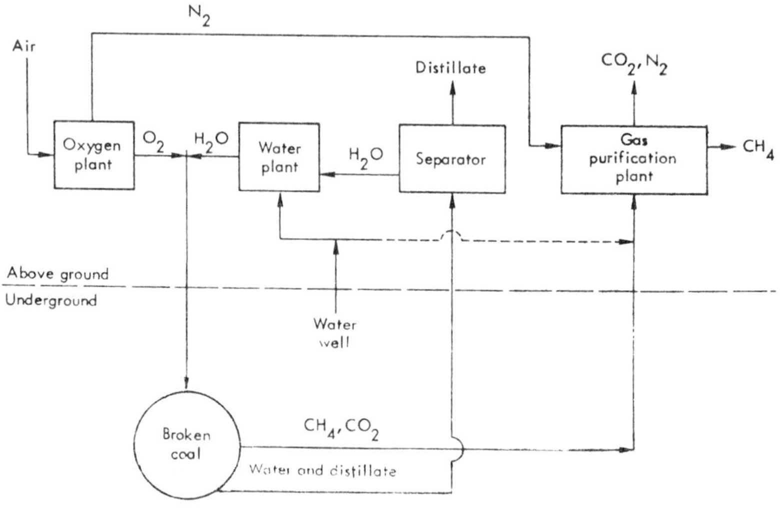

Source: UCRL-51217

The size of the plant would depend on the quantity of gas to be produced, which, in turn, would depend on proximity to a pipeline. It is assumed that 100 BCF per year is the desired rate, that each broken coal unit as shown in the lower part of Figure 5.7 would contain over 5 million metric tons of coal in place and that it would be processed in one year. The consumption and production data are given in the table below.

Annual and Daily Process Rates for a 100-BCF Methane Plant

|  | Annual | Daily |
|---|---|---|
| Gas produced | 100 BCF | 274 MMCF |
| Coal consumed | 5.05 million metric tons | 13.8 thousand metric tons |
| Oxygen consumed | 1.53 million metric tons | 4.19 thousand metric tons |
| Water consumed | 1.54 billion liters | 4.23 million liters |
| Drill holes | ~240      $= 1248\ AF$ | 0.6 |

It appears that the in situ process can be conducted under very favorable kinetic conditions. Figure 5.8 shows a conceptual vertical section through the reacting coal region along with the chemical reactions and approximate temperature distribution. In the inlet region water is being vaporized and heated. As soon as the gases reach the coal, oxygen and the water react very quickly producing the high temperature peak. In the downstream (lower) region carbon monoxide and water react at lower temperature to produce carbon dioxide, methane, and heat. This heat extends the intermediate temperature zone farther and farther from the high-temperature reaction front. Finally, at the successively lower temperatures, water vaporizes and finally condenses.

## FIGURE 5.8: THE CHEMICAL REACTIONS AND THE TEMPERATURE GRADIENT ALONG A VERTICAL SECTION OF THE REACTING COAL BED

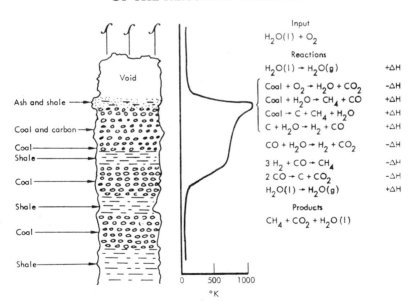

Source: UCRL-51217

Assuming the gas production rates from the table and the area from Figure 5.6, the superficial gas velocity in the broken coal is 3 cm/min, and with reasonable average porosity in the broken coal the actual gas velocity should be 15 to 20 cm/min. Since the reaction zone should be about 10 m thick, the time available for reaction in the high temperature zone is thus probably of the order of one hour.

Data from experiments in the 1100-1300°K range indicate that the reaction kinetics are pseudo-first-order and follow an Arrhenious temperature dependence. Based on the observed rates at the higher temperatures, rates for complete reaction at 700°K are estimated to be of the order of one hour, comparable to the residence times. This rate has not been verified experimentally, however, and it is critical that these kinetics be measured prior to field tests of the process. In any event, the lowest possible temperature should be selected to optimize direct methane production.

## CONCEPT OF PROCESS USING GRAVITATIONAL FORCES AND A SPECIFIED DRILL HOLE PATTERN

### Compensation for Flame-Front Channeling and Liquid Plugging

The seriousness of flame-front channeling, which could result in very low resource utilization, motivated a design for in situ gasification of coal. Also no other process compensated for the phenomenon where the flame front and gas flow are cocurrent (4)-(7). The cocurrent case is important because it is the only one where 100% resource utilization is theoretically possible with a single pass of a flame front. In the cocurrent case, the flame-front propagation is dictated by the full consumption of oxygen and coal, while in the countercurrent case, the front propagation rate is determined by thermal conductivity into the unburned coal, which can leave a large fraction of unburned carbon.

The system is designed to compensate for channeling caused by the spacial distribution of pipes, the inherent "fingering" instability caused by the ash permeability being greater than the coal, and liquid plugging where the liquid source may be either the condensation front that occurs ahead of the flame front, or the general water leakage into the process zone from the surrounding formation.

Two distinct methods, used simultaneously, will cope with flame-front channeling for the elevation and plan views. For the elevation view, gravity and a specially distributed permeability will be employed, whereas for the plan view a well-defined drill hole pattern will be used.

## The Elevation View

A pictorial illustration of the elevation view is presented in Figure 5.9. The first figure shows an early stage of the process just after ignition, and the next illustrates what the process is envisioned to be after it has progressed a fair distance along the coal seam. For this part of the process the design of the system will maximize the use of gravity to move unburned coal into a region where the flame front is beginning to channel. This will have the effect of shutting off the oxygen supply to the leading edge of the channel, thus stopping its progress.

**FIGURE 5.9:  ELEVATION VIEW OF FLAME-FRONT PROPAGATION ALONG A HORIZONTAL COAL SEAM**

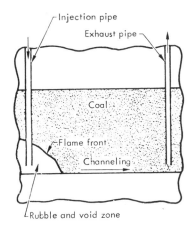

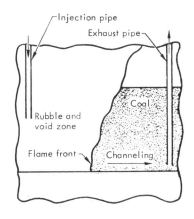

(a)  Early Stage Following Ignition.

(b)  Later Stage Showing Effects of Channeling and Roof Collapse.

Source:  UCRL-51676

To accomplish this it is necessary for the channel to occur at the bottom of the coal seam. It is proposed that this be done in two ways. First the access pipes must extend to be the bottom of the seam, be properly sealed to the seam, and be open to gas flow only at the bottom. This would be sufficient for a uniformly permeable coal bed with minimal liquids since channeling would occur along the shortest distance between the pipes. However, fairly large amounts of liquids are expected to be generated by the thermal cracking of the coal in addition to the large amount of water that frequently occurs in the formations. The liquids will tend to accumulate in the lower portion of the formation, causing a gas flow and thus a flame front override unless steps are taken to prevent it.

Hydraulic or explosive fracturing would enhance preferentially the permeability at the bottom of the seam so that the liquids will drain to the exhaust pipe and be pumped away. A slightly more quantitative description of the requirements for achieving this has been discussed previously (5). The sealing of the pipes to the formation is only important for the exhaust pipe since the gas entry pipe will quickly separate from the coal seam because of the combustion of the coal.

The liquids that are being pumped from the exhaust pipe will contain the pyrolysis products of the coal. Such organic liquids could represent a significant weight fraction of the coal and would be of considerable interest as a petroleum product. The system, therefore, would inherently be a combination liquefaction-gasification process.

This design relies on a definite void being created by the burning of the coal. This places certain restrictions on the properties of the coal or coal ash, as well as on how the process must be carried out. The coal ash content and structural strength must not be so high that it can, even in a partially crushed state, adequately support the coal above it and thus hold open a potential channel. If there is very little ash, or if the crushed ash has a permeability considerably lower than that of the coal, there should be no such problem. In difficult situations where the ash does hold channels open, the severity of the problem could be modified by an appropriate vertical distribution of extraction holes in the exhaust pipe, or by a liquid level control at the exhaust pipe that would raise the liquid level in the seam to shut off bottom channeling as needed. This method does not work in the case where the flame front and gas flow directions are countercurrent. In the countercurrent case, even if the coal were fully combusted in the flame front, which is not likely, the falling of coal into a channeled region would not affect significantly the flame-front propagation.

The explosive fracturing of the coal must satisfy two major requirements. The coal in the upper portion of the bed must be made sufficiently weak so that it can easily fall into the void created at the bottom. This requirement is believed to be far less stringent than requiring a significant permeability enhancement, which is not needed in this case. However, the bottom of the bed must be made highly permeable since the potentially plugging liquids must be drained off at the bottom, and because the gas flow rate must be largest at the bottom to ensure that channeling will begin there. For permeability enhancement at the bottom, explosive fracturing must be done properly. A highly crushed or fractured zone of coal will be far more susceptible to liquid plugging than one where the cracks are more spaced and orientated in the intended direction of the gas flow.

When liquids are injected into a formation there is a tendency for one crack to dominate over all others (8). If a large number of cracks are needed, it may be possible to accomplish this by modifying the injection procedure. The coal could be notched in advance with the desired spacing and direction for the cracks. The injection might then be carried out by isolating different vertical zones and injecting them separately.

How the flame front propagates where unburned coal is discontinuously falling into the front or possibly upstream of the front is difficult to quantify. However, it is expected that the success of the process will be relatively insensitive to such details. If coal falls into a void upstream of the front, the front should then spontaneously shift back upstream, perhaps even in a stepwise manner, and consume that coal before moving on. This only requires that the upstream coal have a mechanism for becoming hot enough to ignite. The zone will have sufficient residual heat from the flame front to ignite the coal. However, if it does not, the coal would then be ignited by the flame front temporarily reversing direction and moving up to burn the coal by the mechanisms responsible for countercurrent burning before proceeding along its original concurrent direction.

### The Plan View

An illustration of the plan view is presented in Figure 5.10. The concept consists of an arbitrarily extendable drill hole pattern that will correct for flame-front channeling in the horizontal direction. A scalloped flame-front profile develops as the flame front moves through the pattern of drill holes because the front always channels into an exhaust hole. Each sequential row of drill holes is positioned so that the scalloped flame-front pattern that forms at one row is inverted as it passes on to the next row. In this way the depth of the pattern is limited and compensated for so that a potential 100% resource utilization is possible.

A possible process sequence is that the coal is first ignited at all the drill holes along row one. Air or oxygen and steam are injected at row one and the combustible product gas exhausted at row two. At some point before

## FIGURE 5.10: THE PLAN VIEW

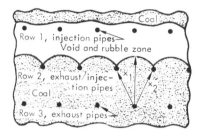
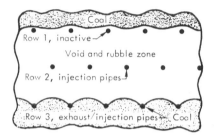

(a) Flame Front Scalloping Upon
    Reaching First Row of Exhaust
    Pipes.

(b) Flame Front Scalloping Upon
    Reaching Second Row of Ex-
    haust Pipes.

Source: UCRL-51676

reaching row two the individual fronts around each hole will coalesce into a single long flame front which is then maintained with a repeating, but continuously changing pattern for the rest of the process. Upon reaching the exhaust pipes at row two, the flame front will have coned forming a scalloped shape as illustrated in Figure 5.10a. At this point the pipes in row two would become injection points and the pipes in row three the exhaust points. If one pipe starts to exhaust from the drill holes at row three before the flame front reaches row two, it will have the effect of flattening out the flame front which might be desirable under some operating conditions. Similarly, one may not stop injecting at row one until after the flame front is well past row two and the effects of roof subsidence at row two have stabilized.

The drill holes for each sequential row must be placed so that the scalloped effect in the flame front keeps inverting as the flame front passes from one row of holes to another. This can be accomplished by ensuring that the gas mobility along the characteristic general path direction described by $x_1$ is greater than that along $x_2$. To do this the distance along $x_1$ is made shorter than $x_2$ if the permeabilities along both paths are equal. However, it is possible to allow $x_1$ to be greater than $x_2$ and still obtain the desired effect if the permeability along $x_1$ is adjusted so that it is appropriately larger than that along $x_2$.

Some coal could be left behind as pillars to support the roof by reversing the condition just described. That is, let the gas mobility along path $x_2$ exceed that along path $x_1$, or let the distance along $x_2$ be shorter than that along $x_1$. If this is done, pillars of coal will be left behind in an obvious manner. The exact size of the pillars will depend on the magnitude of the pipe spacings and gas mobilities along the different paths. However, this approach is generally undesirable from a resource utilization standpoint and would be applicable only in some special circumstances.

When explosive fracturing is used, it is envisioned that the new holes drilled ahead of the flame front would be filled or appropriately injected with explosives as described previously and detonated when the flame front reaches the proper distance from the holes. One favorable feature of the explosive fracturing procedure is that there will be a large amount of free surface, as defined by the flame front, near the explosives. Also the nonuniform radial distribution of permeability that will naturally occur around each exhaust hole that has been explosively fractured is not detrimental to the process. In fact it is likely to be very beneficial in that the permeability will increase as the flow lines converge into the exhaust pipe. This will greatly reduce the overall flow resistance that is always highest around the access pipes in in situ processes because of the diverging or converging of the flow. It is not certain that this would always be the optimum mode of operation since in some situations there might be good reason to use separate holes for explosive fracturing.

A phenomenon of critical concern is the manner in which the roof collapses. If the coal is completely removed over a broad front as described there is no question as to whether the roof will collapse; however, it is uncertain as to exactly how its collapse will affect the process. The guaranteed collapse could favorably assist this method by preventing channeling in the elevation view. However, there is the danger of forming highly permeable cracks either all the way to the surface or to other formation layers. The primary effect of such cracks leading away from the process zone would be to cause a severe gas loss if the cracks are large.

An additional difficulty characteristic of this particular process is that there is the potential that the gas leakage rate will be accumulative over the entire region that has been processed previously. Also, because of the nature of cocurrent flame fronts, the high pressure side of the reaction is associated with the ash were one would expect the maximum gas leakage problems. This potential difficulty might require that formation plugging materials such as muds, foams, or even water be injected down the holes in the rows that are no longer being used for the process. In this manner one would hope to achieve a sealed plug that follows some distance behind the flame front, but not so close as to interfere with the active gas injection wells.

## PROPRIETARY PROCESSES

### Controlled Rate of Combustion

U.S. Patent 3,997,005 describes a method for in situ gasification by penetrating the coal bed with a group of boreholes arranged in general alignment and spaced apart from one another along a plane perpendicular to the plane of maximum permeability, and penetrating the coal bed with a second group of spaced-apart boreholes arranged generally parallel to the first boreholes.

After initiating combustion in the coal bed the first boreholes are interconnected to form a single combustion zone in registry with all of the first boreholes. The combustion-supporting gas is introduced into the boreholes at a rate sufficient to support and maintain the combustion zone. Gaseous products of combustion are withdrawn from the combustion zone through the second boreholes. The flow rate of gaseous combustion products is withdrawn through each borehole of the second group controlled to selectively vary the pressure drop of the combustion products flowing between the combustion zone and each of the second boreholes. This provides and maintains uniform propagation of the combustion zone towards the second boreholes.

### Increasing Air Injection Rates by Gas Fracturing

U.S. Patent 3,775,073 describes a process whereby two or more wells are completed within a coal deposit. A radially extended horizontal fracture is induced through the coal deposit so as to interconnect the wells.

The coal deposit is subjected to an excess of a first combustion-supporting gas at a pressure greater than the overburden pressure so as to distribute the combustion-supporting gas throughout the coal network. Subsequently, the coal deposit and the first combustion-supporting gas are ignited while simultaneously preventing any fluid or gas production from the coal deposit, so as to form a network of crumbled coal within the coal deposit. Continued injection of a second combustion-supporting gas into one or more wells and production of combustible gas and coal tar liquids from one or more of the wells completes the process.

### Addition of Carbon Dioxide and Steam to Injection Well

U.S. Patent 3,770,398 describes a process whereby a coal deposit is burned to raise the temperature therein above about 1000°C as steam is injected to produce a water-gas shift reaction product gas. The process comprises introducing carbon dioxide in the coal deposit in order to favor the reaction kinetics to the water-gas shift reaction. The combustion and steam injection steps may be accomplished simultaneously or in separate phases wherein the temperature of the reservoir is maintained at/or above about 1000°C. The carbon dioxide introduced is generally obtained from the gas produced from the coal deposit. The steam introduced may be contacted with the produced gas from the coal deposit to regain the waste heat content so as the further expedite the energy balance of the system.

## Introduction of Superheated Steam

U.S. Patent 3,734,184 describes a method for producing volatile hydrocarbons from a rubblized zone of coal through the initial injection of a superheated steam stream, with subsequent combustion of the coal zone, introduction of water in the burned zone and subsequent production of steam for the sustenance of the superheated steam production for continued volatile hydrocarbon removal.

The process comprises completing one or more wells within a coal deposit and rubblizing the coal deposit above the well. Subsequent to rubblization, superheated steam is introduced into the rubblized coal deposit through the well with subsequent production of volatized hydrocarbons and a synthetic gas, having a high calorific energy value, from the rubblized coal deposit, through the well. Subsequent to hydrocarbon production, the rubblized coal deposit is burned and water introduced into the burned rubblized coal deposit in order to form a superheated steam.

It is preferred that the superheated steam have a temperature of from about 1200 to 2000°F with the rubblized coal deposit being burned to raise the temperature therein from about 1200 to 2000°F. Generally, it is preferred that at least two wells are completed within the coal deposit, with one well being utilized as an injection well and the other as a production well. The process of production of volatile hydrocarbons, burning, and introduction of water to form superheated steam may be repeated until no further depletion of hydrocarbon or synthetized gas production is achieved. It is also preferred that the superheated steam produced from the burned rubblized coal deposit be produced and injected into another rubblized coal deposit from which volatile hydrocarbons and synthetic gas may be produced.

## Regulation of Fracture Network Width by Injection Pressure Control

In typical in situ gasification two or more wells are completed into the coal seam and all sections therewith are segregated from the wellbore by isolation means so that only the coal seam is left exposed and in direct contact with the wellbore. A horizontal fracture is created through the coal seam so as to connect the respective wellbores of the completed wells. The coal seam is ignited through one or more of the wells and the combustion front propagated from one or more of these injection wells to the production wells by injection of a combustion-supporting gas into the ignited wells.

The amount of combustion movement is directly proportional to the oxygen injection rate so that the advancement of the combustion front may be controlled as it moves throughout the reservoir.

It is common practice to separate a fractured network by the use of propping agents. In the combustion of a coal seam, however, it has been found that air injection into the coal bed, which has been hydraulically fractured and heated, results in the production of hot coal tar products which flow into the fracture and when cooled ahead of the combustion front become highly viscous or solid and plug the fracture. Any propping agent contained with this fracture will tend to coagulate the viscous fluid and further block the fracture thereby preventing further gas injection. In U.S. Patent 3,628,929 it was found that by injecting air or other combustion-supporting gas mixtures at a pressure sufficient to raise the overburden strata this problem is overcome. The unobstructed path of flow afforded the injected gas as the overburden is raised allows a greater amount of gas to be introduced into the coal seam and greater quantities of coal to be contacted and consumed.

The process is then self-regulating in that the overburden can be lifted or lowered even to the point of closure if necessary to provide the liquid coal tar, flammable gas and injection gas flow rates. The continued production of a flammable gas and coal tar liquids is thereby afforded.

## Underground Movable Furnace Wall

U.S. Patent 3,497,335 describes a process whereby gasification of coal underground is accomplished using a movable furnace wall structure spaced from and substantially parallel to the coal face. This forms an enclosed furnace space to which oxidant is supplied, preferably via a multiplicity of openings through the wall, to promote gasification of the coal along this face. The hot products of combustion are extracted continuously from the furnace space and the wall structure is advanced as the coal face is consumed, preferably by connection with remotely controlled self-advancing roof supports.

## Omission of Combustion Section

In gasification, during combustion coal undergoes a volume reduction in the preheating zone by the expulsion of volatile components, particularly in the case of gas-rich coal, as a result of which a widely branched and very intensive formation of fissures is produced, which alone are sufficient for the conduction of the gasification agent.

In the utilization of this phenomenon, the gasification process according to German Patent 1,015,982 is carried out in such a manner that a combustion section is omitted for the gasification procedure, the roof is lowered by appropriate means so that no combustion section can form, and the gasification agent reacts only with the coal in the cracks appearing in the seam from the heat. For this purpose a flow channel is made parallel to the connecting section between the air inlet and the gas removal, by horizontal boreholes or by an additional section, only for the introduction of the gasification process.

Before ignition of the coal for gasification, the roof in the connecting section is brought down by blasting, and thus the rock above the connecting section is destroyed and the latter is filled with rubble, so that further spreading of the combustion via the completely decarburized connecting section is no longer possible.

## Borehole Heating Process

German Patent 1,015,181 describes a process for energy production by means of underground gasification of bituminous deposits especially low-grade ones, in which thermal energy is supplied to the deposits intersected by boreholes for gas production from bituminous rock masses. The process makes use of the fact that heating devices, such as electric heating pipes, are installed in the boreholes and that by their controllable energy release the energy necessary for the gas production is covered and the gas production is thus influenced.

According to a preferred form the bituminous deposits are both degassed and, by means of the inherent moisture of the deposits or by means of added gasification agents such as water or steam, gasified. The temperature of the pit itself can be used as energy additional to the heating devices for the gas production. The process thus abstains from burning the intersected bituminous deposits in order to obtain the energy necessary for the gas production by this combustion, but uses external energy to cover the energy requirements. By these measures it is possible to produce a gas with a high calorific value and of adjustable composition. In particular, it is possible to remove from low-grade deposits a gas with a high calorific value, which can be used, for example, for the operation of gas turbines.

The energy production obtained in this way by direct transfer of the gas produced into machines such as gas turbogenerators is essentially higher than the consumption of energy which has to be fed into the heating devices of the boreholes to cover the energy for the gas production. The latter is particularly true if cheap off-peak electricity is used for gas production in the seams, and also if the process is carried out in bituminous deposits of great and very great depth, where the rock pressure and above all the rock temperature can support the degassing process. Thus, large amounts of gas can be removed from deep seams after only slight heating. It is also possible to carry out the gas production under pressure, as a pressure gasification, the pressure being chosen in accordance with the rock pressure.

## Utilization of Heat Losses

It has been reported that during the gasification of a horizontal coal seam, measurements of the heat in the roof layers indicated that after several months the roof layers still showed temperatures of 550°C at 5 m above the combustion zone, and that 19 m above the combustion zone the temperatures in the rock were still above the boiling point of water.

The process described in German Patent 952,839 is characterized in that in the presence of several superimposed seams a lower seam is gasified according to conventional underground gasification processes, and at the same time by utilizing the heat transferred to the roof of this seam, the next higher seam is carbonized and the carbonization gases produced are removed. Furthermore, the process relates to gasification of the fuel residue (coke) in the upper seam after burning out of the lower seam, according to conventional underground gasification processes.

Although the heat loss to the country rock will always be different, the heat will in any case be sufficient to make the carbonization process in the seam next above the one to be gasified possible. The carbonization gases formed in the upper seam are of essentially higher quality than the gases from the gasification process in the lower seam, because the carbonization gases in the upper seam do not contain the nitrogen fraction of the combustion air. The carbonization gases are obtained in the process as by-products of the actual underground gasification process.

REFERENCES

(1) H.V. Sears, U.S. Patent 3,563,606, February 16, 1971, *Method for In-Situ Utilization of Fuels by Combustion;* assigned to St. Joe Minerals Corp.
(2) A. Gibb, et. al., *Underground Gasification of Coal;* Pitman and Sons, Ltd, London 1964, 205 pp.
(3) H. C. Carpenter, H. W. Sohns, and G. U. Dineen, "Oil Shale Research Related to Proposed Nuclear Projects," in *Proc. Engineering with Nuclear Explosives,* CONF-700101 (1970) p. 1364.
(4) D. W. Gregg, *The Stability of Flame Front Propagation in Porous Media with Special Application to In-Situ Processing of Coal,* Lawrence Livermore Laboratory, Rept. UCRL-51595 (1974).
(5) D. W. Gregg, *Liquid Plugging in In-Situ Coal Gasification Processes,* Lawrence Livermore Laboratory.
(6) C. A. Komar, W. K. Overbey, and J. Pasini, *Directional Properties of Coal and Their Utilization in Underground Gasification Experiments,* Bureau of Mines Advancing Energy Utilization Program, Technical Progress Rept. 73 (1973).
(7) A. D. Little, *Report to the U.S. Bureau of Mines, A Current Appraisal of Underground Coal Gasification,* Rept. C-73671 (1971).
(8) P. P. Klimentov, *Izv. Vyssh. Ucheb. Zaved. Geol. i Razved.* (10), 97-105 (1964).

# Operational Techniques Designs and Studies

The material in this chapter was excerpted from MERC/SP-75/1, PB 209 274, PB 241 892, TID 26825, UCRL-50026-75-1, UCRL-50026-75-2, UCRL-50026-75-3, UCRL-50026-75-4, UCRL-51790. For a complete bibliography, see p 251.

## SITE SELECTION

The total amount of coal in the United States is estimated to be 3200 billion tons (1). Approximately half of this coal, about 1560 billion tons, is considered to be an identified or demonstrated resource (1)(2)(3). This includes beds of bituminous coal and anthracite 14 in. (36 cm) or more in thickness and beds of subbituminous coal and lignite 30 in. (76 cm) or more in thickness at depths from 0 to 3000 ft (900 m). All of this coal except that at very shallow depths and the anthracite (2) could be suitable for in situ exploitation. The demonstrated resources of anthracite and semianthracite are 13 billion tons. The remaining strippable resources of coal, generally at depths less than 120 ft (36 m), are 117 billion tons (4). Thus, the demonstrated resources which are possibly suitable for in situ exploitation amount to 1430 billion tons. This includes 1000 billion tons of bituminous and subbituminous coals and 430 billion tons of lignite.

### Distribution of Coal Resources by Location

The coal fields in the United States are concentrated in several areas, as previously shown in Figure 1.2. These fields are geographically subdivided by the U.S. Geological Survey into seven provinces and subregions. Of the bituminous and subbituminous coals which could be exploited by in situ methods, 86% are contained in the Eastern Province (235 billion tons), the Eastern Interior Province (181 billion tons), and the

94

Western and Rocky Mountains Provinces (400 billion tons). The lignite which could possibly be exploited by in situ methods is essentially all contained in the Northern Great Plains Province (427 billion tons). These five important areas are discussed below.

The Eastern Province: Bitiminous coal occurs throughout the Appalachian region in the Eastern Province in formations of Pennsylvanian and Permian ages. The bituminous coal ranges in rank from low-volatile to high-volatile C coals. The sulfur content ranges from about 0.5% to over 6%.

The bituminous coal is contained in approximately 90 minable seams throughout the Appalachian region. The major seams are continuous over large areas. The attitudes of the seams vary from flat to gently dipping, with local steepening in some fields. The seams pinch out and swell and are sometimes faulted or interrupted by erosion. About 2/3 of the seams can be strip mined to various extents, depending on location and depth of overburden.

The Interior Province—Eastern Region: The coal in the eastern Interior Province occurs in an oval basin covering parts of Illinois, Indiana, and western Kentucky. Present strip mining is essentially restricted to the margins. The beds are horizontal to gently dipping and Pennsylvanian in age. The coal is low-volatile to high-volatile C bituminous and is high in ash and sulfur and generally unsuitable for coking.

The Interior Province—Western Region: The coal seams in the Western Interior Province occur in formations of Pennsylvanian age. Although the bituminous coal seams are essentially flat to gently dipping, there is little correlation of seams from field to field. The seams are discontinuous and irregular in thickness, thinning and swelling in comparatively short distances. The bituminous coal ranges from low-volatile to high-volatile C in rank, generally from medium to high in sulfur and from low to high in ash.

The Rocky Mountain Province: Coal beds in the Rocky Mountain Province occur in formations of Tertiary and Cretaceous ages. The seams are generally flat to gently dipping, but local steepening is present in some areas. There is a great variation in continuity and thickness, with seams as thick as 90 ft (27 m).

The Northern Great Plains Province: The coal deposits of the Northern Great Plains Province include mostly subbituminous coal and lignite in formations of Tertiary and Cretaceous ages. The subbituminous coal seams range up to 220 ft (67 m) in thickness, and the lignite seams range up to 40 ft (12 m) in thickness.

### Distribution of Coal Resources by Seam Thickness

Coal resources are calculated by beds in categories of thickness according to standard procedures of the U.S. Geological Survey. The ranges of thickness of these categories for different ranks of coal and the percentages of resources contained within each category have been reported as in Table 6.1 (2).

### TABLE 6.1: USGS CATEGORIZATION OF COAL RESOURCES BY SEAM THICKNESS

| Category | Thickness for anthracite and bituminous coals (in.) | Thickness for subbituminous coals and lignite (ft) | Fraction of total U.S. resources (%) |
|---|---|---|---|
| Thin | 14 - 28 | 2.5 - 5 | 44 |
| Intermediate | 28 - 42 | 5 - 10 | 26 |
| Thick | > 42 | > 10 | 30 |

Source: TID-26825

This particular categorical description of coal resources by thickness is not useful for determining the quantities of coal which could be suitable for in situ processing by horizontal burning ("thin"-seam methods) or by vertical burning ("thick"-seam methods). For this purpose, the "thin" category should include coal seams up to, say, 15 ft thick, and the "thick" category should include seams over, say, 30 ft thick.

The coal resources of the U.S. can easily be divided into these categories if it is assumed that the resources of each state are contained in seams of the average thickness of seams mined in that state (5). However, more information is available on shallow coal seams suitable for stripping (4). From this information, the strippable coal resources may be divided into thickness categories by the amount of coal contained in each seam in each state and the average seam thickness. The results of each of these methods are given in Table 6.2.

TABLE 6.2: SEAM-THICKNESS CATEGORIES USEFUL FOR IN-SITU
PROCESSING CONSIDERATIONS

| Category | Thickness (ft) | Fraction of total U. S. resources[a] (%) | Fraction of total remaining strippable resources[b] (%) |
|---|---|---|---|
| Thin | 0 - 15 | 82 | 64 |
| Intermediate | 15 - 30 | < 1 | 9 |
| Thick | > 30 | 16 | 18 |
| Total | | 98[c] | 91[c] |

[a]From the coal resources of each state (1) and the average thickness of the seams mined in that state (5)
[b]From the remaining strippable coal resources in each seam of each state and the average thickness of each seam (4)
[c]The totals are less than 100% because complete information is not available for each state

Source: TID-26825

A detailed distribution of strippable resources by thickness is presented in Figure 6.1. Neither of these methods necessarily provides an accurate distribution by seam thickness of the coal resources possibly suitable for in situ exploitation because of obvious reasons. However, it is possible that the distribution derived by either method could be applicable for this purpose. The results are fairly consistent, indicating that more than 60% of the resources are contained in seams less than 15 ft thick and less than 20% of the reserves are contained in seams greater than 30 ft thick.

**Assessment**

There are significant deposits of coal in several widespread areas of the United States which could be suitable for in situ exploitation using thin-seam and thick-seam methods. There is about four times as much coal in thin seams as there is in thick seams. The areas suitable for thin-seam methods include the Eastern Province, the Eastern region of the Interior Province, the Northern Great Plains Province, the Rocky Mountains Province, and possibly the Western region of the Interior Province. Significant coal deposits which would be suitable for in situ exploitation using thick-seam methods are restriced to fewer, more remote areas, such as Montana and Wyoming.

**FIGURE 6.1: DISTRIBUTION OF STRIPPABLE COAL RESOURCES
BY SEAM THICKNESS**

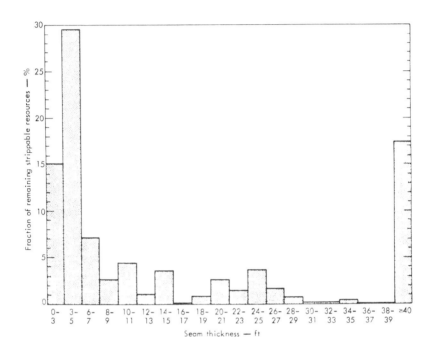

Source: TID-26825

LLL Exploratory Drilling: Coal deposits that appear suitable for in situ gasification using the LLL scheme were identified in six western states and Alaska. However, the best sites for the experimental program are in the Powder River Basin of Wyoming and Montana, and in southwestern Wyoming.

Initial exploratory drilling was carried out from December 1974 to March 1975. Five sites on public land in the Powder River Basin of Wyoming and Montana and one site on Kemmerer Coal Company land near Kemmerer in the Hams Fork coal region of Wyoming were investigated. Deep thick coal seams were identified at each of these sites (Figure 6.2). Hole conditions prevented obtaining open hole logs through the major coal seam at both the Hay Creek and Hoe Creek sites. However, it was determined that the coal was more than 15 m thick at each site. Pinette Draw was selected from the Powder River Basin sites for detailed characterization of a deep, thick coal. The comparative ease of drilling, the lack of major drilling-fluid losses within the coal beds, and the paucity of water-bearing sands, all contributed to this selection.

Of the four holes drilled through the Adaville coal seams southwest of Kemmerer, Wyoming, three encountered seams more than 15 m thick.

## FIGURE 6.2:  LOCATION OF COAL SITES INVESTIGATED BY LLL

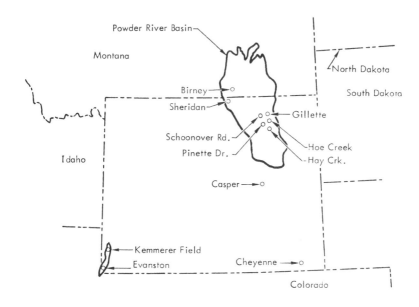

Source:  UCRL-50026-75-1

## SITE CHARACTERIZATION

The following description is an illustrative example of what is required to characterize a site adequately. This work was performed by LLL after site selection. LLL conducted a field characterization program at the Hoe Creek site in preparation for fracturing and gasification experiments.

Figure 6.3 shows the relative locations of the main areas of activity, which were Site 1 (Hoe Creek Experiment No. 1), Site 2 (where extensive hydrologic tests were conducted), and Site 3, the location of the proposed five-spot gasification experiment. Additional exploratory holes were drilled deeper in other areas to characterize the strata down to and including the Wyodak-Anderson coal seam.

**FIGURE 6.3: HOE CREEK SITE, CAMPBELL COUNTY, WYOMING**

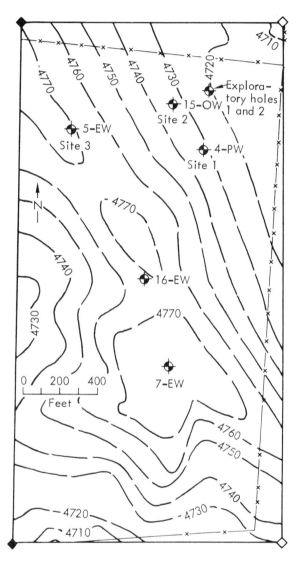

*NOTE:   The contour interval is 10 ft.

⊕        indicates a drill hole.

×        indicates a fence line.

Source: UCRL-50026-75-3

Extensive geophysical logs and core samples were obtained in some holes, while engineering survey logs were taken in all holes that were drilled and completed.

In addition to the main tasks of characterizing site geology and hydrology, they also sampled and analyzed water and tested a method for high-frequency electromagnetic tracking of in situ fluid movement.

## Geology

The Hoe Creek was initially characterized from two exploratory holes that were drilled in the winter of 1975. The Hoe Creek Nos. 1 and 2 holes were drilled to a total depth of about 900 ft, and from them it was determined that 94 ft of the Wyodak-Anderson seam occurred at 768 ft. Major drilling problems were encountered, causing initial abandonment of the site as a location for deep gasification. The presence of the Felix coal seams at shallow depths, however, made the site a possible location for a shallow gasification experiment, and additional holes were drilled to detail the geology and hydrology of the Felix coal seams.

The Felix No. 2 coal seam is nearly flat and a constant 25 ft thick over the site. No clay partings were found, although a 2-ft-thick high-ash zone shows on these logs in the upper part of the seam. The Felix No. 2 seam is overlain by 14 to 18 ft of very-fine-grained sandstone and siltstone. The 38 ft of separation found at the exploration site is a local anomaly. The overlying Felix No. 1 seam is 10 to 11 ft thick and free of clay partings. From the top of the Felix No. 1 seam to the surface is a variable sequence of shale, siltstone, and sandstone. Table 6.3 shows typical coal analyses for the Felix No. 1 and 2 seams.

While characterizing the site, another attempt to drill through the deeper Wyodak-Anderson coal was made. Drilling breaks and sparse cutting returns indicated that the Wyodak-Anderson seam was 95 ft thick and 795 ft deep.

## Test-Well Design and Development

Figure 6.3 shows a typical observation well. The depth of the well components varies with site location. In general the slotted polyvinyl chloride (PVC) casing is adjacent to the seam of interest, which in most cases is Felix No. 2. In resistivity wells (for [High-Frequency Electromagnetic] HFEM) measurement and dewatering wells, an additional section below Felix No. 2 is added for instrumentation or pumps.

## TABLE 6.3: ANALYSIS OF COAL FROM THE FELIX Nos. 1 AND 2 SEAMS

| | Proximate analysis (%) | | | Ultimate analysis (%) | |
|---|---|---|---|---|---|
| | As received | Dry basis | | As received | Dry basis |
| **Felix No. 1 seam** | | | | | |
| Moisture | 31.88 | – | Moisture | 31.88 | – |
| Ash | 4.36 | 6.40 | Carbon | 46.89 | 68.84 |
| Volatile | 31.38 | 46.06 | Hydrogen | 3.56 | 5.22 |
| Fixed carbon | 32.38 | 47.54 | Nitrogen | 1.16 | 1.70 |
| | 100.00 | 100.00 | Chlorine | 0.01 | 0.01 |
| | | | Sulfur | 0.46 | 0.68 |
| Btu | 8093 | 11881 | Ash | 4.36 | 6.40 |
| Sulfur | 0.46 | 0.68 | Oxygen (diff) | 11.68 | 17.15 |
| | | | | 100.00 | 100.00 |
| **Felix No. 2 seam** | | | | | |
| Moisture | 30.11 | – | Moisture | 30.11 | – |
| Ash | 4.05 | 5.80 | Carbon | 48.38 | 69.23 |
| Volatile | 32.12 | 45.96 | Hydrogen | 3.66 | 5.24 |
| Fixed carbon | 33.72 | 48.24 | Nitrogen | 1.01 | 1.45 |
| | 100.00 | 100.00 | Chlorine | 0.00 | 0.00 |
| | | | Sulfur | 0.34 | 0.49 |
| Btu | 8359 | 11960 | Ash | 4.05 | 5.80 |
| Sulfur | 0.34 | 0.49 | Oxygen (diff) | 12.45 | 17.79 |
| | | | | 100.00 | 100.00 |

Source: UCRL-50026-75-3

FIGURE 6.3: SCHEMATIC OF A TYPICAL TEST WELL

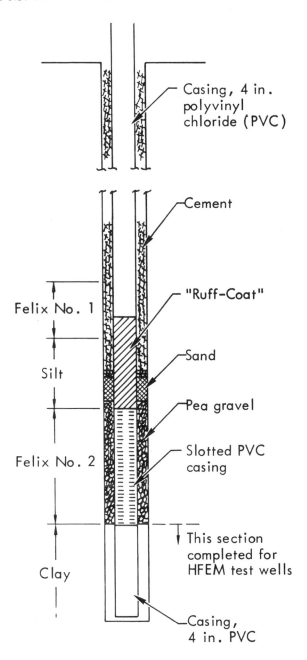

Source: UCRL-50026-75-3

It was necessary to air-lift water from the bottom of the wells to clean and develop them prior to installing the pumps. Pumps had been installed on three occasions with their inlets below the tops of screens in wells, and on all three occasions the pumps clogged with fine sand and silt. Pumps in dewatering wells must be set below the top of the screens. Hence, to assure a reasonable probability that these pumps continue to operate after installation, it is necessary to:

1. assure that they are resistant to sand and silt,
2. take special care in selecting screen slot size, and
3. make sure that well development is effective and complete.

### Hydrology and Permeability

Hydraulic tests were conducted at three sites, namely Sites 1, 2, and 3. It was found that hydraulic-head potential decreases with depth at the Hoe Creek site. Groundwater flow is therefore inferred to have a vertical downward component. Although the water table (top of the saturated zone) had not been definitely located, it was believed it may be about 15 m below the surface at the characterization site and about 27 m below land surface on the ridge above the characterization site.

Table 6.4 summarizes the results of hydraulic tests at Hoe Creek. The slug tests are secondary tests and are included for completeness. They are single-well tests and are influenced by well bore and other close-in effects.

### Joint HFEM-Hydrology Tests

Three separate high-frequency electromagnetic (HFEM) measurements were made during water injection at Site 1. One HFEM measurement was performed while injecting water into the Felix No. 1 seam; the other two while injecting water into the Felix No.2 seam. During the second injection into Felix No. 2, water levels were measured in three observation wells and the injection well.

These experiments demonstrate the feasibility of joint HFEM permeability measurements.

### Water Samples and Analyses

Pumped water samples were collected from various wells at Site 1 and analyzed promptly in the field. The general procedure was to fully develop the well by inducing flow by air lift and submersible pump. After

## TABLE 6.4: RESULTS OF HYDRAULIC AND PERMEABILITY TESTS AT HOE CREEK

| Site No. | Seam | Test wells | Horizontal permeability (D) | Remarks |
|---|---|---|---|---|
| 1 | Felix 2 | 4,8,9,13 | 0.375 (max) <br> 0.197 (min) <br> (Storage coefficient = $1 \times 10^{-3}$) | Drawdown tests, $K_{max}$ direction. N 59° E. $K_{min} \perp K_{max}$ |
| 1 | Felix 2 | 8 | 0.052 | Slug-injection test |
| 1 | Felix 2 | 9 | 0.637 | Slug-injection test |
| 1 | Felix 2 | 13 | 0.187 | Slug-injection test |
| 1 | Felix 2 | 4,8,9,13 | 0.080 (av) | Steady-state and injection test |
| 1 | Felix 1 | 12 | 0.630 | Slug-injection test |
| 1 | Silt (above Felix 1) | 12 | (substantial vertical leakage) | Drawdown test |
| 1 | Silt (between Felix 1 & 2) | 10 | 0.080 | Slug-injection test |
| 1 | Silt (between Felix 1 & 2) | 10,13 | 0.0016 (vertical) <br> (Storage coefficient = $1 \times 10^{-3}$) | Drawdown and consolidation tests of silt samples |
| 1 | Clay (below Felix 2) | 11 | $K \ll K_{silt}$ | From very slow water-level response |
| 2 | Felix 2 | 14,15 | $K_{vert} > K_{horiz}$ | Drawdown test |
| 3 | Felix 2 | DW 1, Res. 1, Perm 1,2. | Both the magnitude and direction of max and min permeability are similar to those evaluated at Site 1. | Drawdown tests |
| 3 | Silt (between Felix 1 & 2) | DW 1, Res. 1, Perm 1,2. | ~0.20 (vertical) | Drawdown tests |

Source: UCRL-500-26-75-3

the wells were developed, pumped samples were collected. Table 6.5 summarizes the results of several analyses.

### TABLE 6.5: ANALYSES OF SUBSURFACE WATERS AT THE HOE CREEK GASIFICATION SITE

| | Analysis[a,b] | | |
| --- | --- | --- | --- |
| | 1 | 2 | 3 |
| Bicarbonate (as $CaCO_3$) | 288 | 280 | 481 ± 3 |
| Carbonate (as $CaCO_3$) | 0 | 0 | 0 |
| Carbon dioxide ($CO_2$) | 48 | 40 | 27 ± 7 |
| Chloride (Cl) | 11 | 13 | 12.8 ± 0.6 |
| Hardness | | | |
|   Calcium (as $CaCO_3$) | 460 | 690 | 75 ± 10 |
|   Total (as $CaCO_3$) | 750 | 1075 | 107 ± 22 |
| Nitrate ($NO_3$) | 0.3 | 0.3 | 0.6 ± 0.2 |
| Oxygen, dissolved ($O_2$) | 3.0 | 4.5 | 4.4 ± 1.7 |
| Sulfate ($SO_4$) | 750 | 1100 | 152 ± 37 |
| Total dissolved solids | 1882 | 2738 | 1002 ± 34 |
| pH | 7.3 | 7.5 | 7.8 ± 0.1 |
| Specific conductance (mS/m) | 192 | 290 | 102 ± 2 |
| Color | 40 | 80 | 83 ± 69 |
| Turbidity | 10 | 18 | 24 ± 17 |

[a]All concentrations in mg/l
[b]Analyses are as follows: (1) Felix No. 1, (2) Silt between Felix No. 1 and No. 2, (3) Felix No. 2

Source: UCRL-50026-75-3

## COAL PROCESSING EXPERIMENTS

### High Pressure Experiments

The purposes of the high pressure, coal-combustion experiments are to:
1. Examine the properties of a moving, reaction zone in a packed bed of coal under operating conditions anticipated for in situ gasification.

2. Study the chemical reactions and product composition as a function of variable feed and operating conditions.
3. Provide experimental data for coal-processing calculational models that include combined heat and mass transport, flow, and multiple chemical reactions.

In one such experiment, Wyodak coal was reacted in the 0.3-meter reactor with steam and oxygen at constant input rates of $9.1 \times 10^{-3}$ and $5.0 \times 10^{-3}$ g·mol/s, respectively. The direction of burn and flow was downward (co-current burn), and the bed pressure was held at 0.81 MPa (117 psi). The temperature rises as the reaction zone approaches, reaches a peak, and then falls off as the char is consumed. The plateau corresponds to the temperature of the mixture of steam and product gas that precedes the reaction zone.

Figure 6.5 shows the temperature profiles as the reaction proceeds through the bed. The profiles continue to change shape due to a combination of heat losses and the fact that the width of the reaction zone is of the same order as the total bed length.

**FIGURE 6.5: TEMPERATURE PROFILES AS THE REACTION PROCEEDS THROUGH THE PACKED BED**

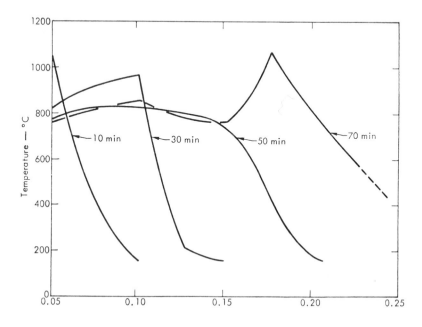

Source: UCRL-50026-75-1

The gas produced in this experiment had a heating value of 9.4 mJ/m³ (252Btu/ft³). Liquid products were 96% water and 4% tar. The tar fraction was ~3 wt% of the coal. A reasonable overall and elemental mass balance were obtained. Assuming 18.6 mJ/kg (8000 Btu/lb) heating value for coal, the conversion energy distribution was approximately 33% in the product gases, 36% in the unburned char, 20% in unrecovered heat losses, and the remainder in the tar and in product sensible heat.

The product gas composition was fairly constant when the reactant flow rates were held steady. However, in an earlier experiment, the input steam rate was increased during the course of the run, while the oxygen rate was held constant. Figure 6.6 shows the change in product gas composition under these conditions. Qualitatively, the increased hydrogen and carbon monoxide production is consistent with the occurrence of the steam-carbon reaction.

### FIGURE 6.6: THE EFFECT OF THE INPUT-STEAM FLOW ON PRODUCT-GAS COMPOSITION IN THE HIGH PRESSURE REACTOR

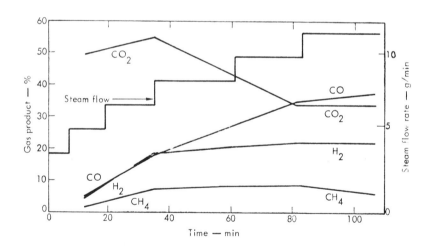

Source: UCRL-50026-75-1

Apparent conclusions are:

  1. Heat losses by conduction/radiation are significant. Combustion temperatures are controlled by heat losses and by input operating parameters.

2. The thermal profile is fairly well defined, but continually changes shape as the reaction proceeds through the bed.

3. A significant radial temperature gradient exists across the reactor bed.

4. Tar plugging is not a problem, but channeling does become pronounced as the bed is consumed.

5. Overall operation is fairly uniform with time. Therefore, manual control is presently adequate, and gas analysis every 10 to 15 min (which is the current turnaround time for the in-line gas chromatograph) is sufficient.

6. Heat and mass balances need to be improved by tightening mass inventory—i.e., reactant and product flow rates, residue, tar and water.

7. The width of the combustion zone needs to be narrowed by the use of smaller particles.

## Quartz Tube Combustion Experiment

A silica-glass combustion tube was used to allow observation of the principal features of the gasification process. One can clearly see coal drying, tar production, hydrocarbon mist production, char formation, char combustion, and ash formation. The combustion tube is 6.6 cm in diam and 1 m long. The tube is packed with 1.7 kg of subbituminous coal. The resulting coal bed has a porosity between 40 and 50%. When burning in a downward, cocurrent mode, the maximum burn-front temperature, burn-rate, and outlet gas composition were measured as a function of air flow rate.

The gas composition at one location in the combustion tube during the transit of the burn front is given in Figure 6.7 together with the temperature at the point where the gas was sampled. The lack of hydrogen and methane at the thermal maximum tends to confirm the idea that by the time the burn front reaches a certain place in the coal bed, no coal actually remains. The coal ($\sim CH_{0.9}$) has already been pyrolyzed, without oxidation, to char ($\sim CH_{0.3}$). It is mostly carbon that is actually burning to produce carbon dioxide and carbon monoxide.

When coal is burned in the countercurrent mode (air flow downward and burning upward), the process proceeds as shown in Figure 6.8. Char is produced during the upward sweep of the flame, and then subsequently burned during the downward passage of the flame.

Liquid hydrocarbons precede the burn front during cocurrent burning.

**FIGURE 6.7: GAS COMPOSITION AT ONE POINT IN THE
COMBUSTION TUBE DURING COCURRENT BURN**

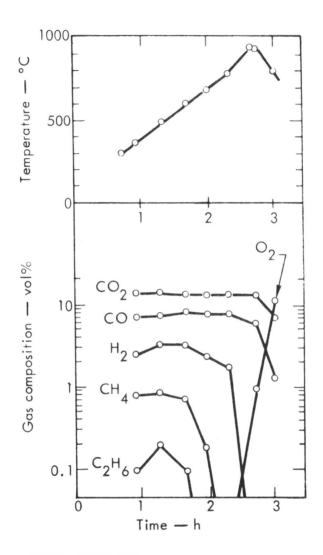

Source: UCRL—50026-75-1

This "tar front" does not become as massive as one would expect based
on the amount of tar produced in pyrolysis experiments. Apparently
some of the tar breaks down into char and lighter liquids or gas due to the
repeated heating it undergoes as it is driven along by the advancing burn.

This char deposition by tar is important since it may be a process whereby a burn would continue across a barren sandstone or shale layer in a coal bed.

**FIGURE 6.8: SCHEMATIC REPRESENTATION OF COUNTER-CURRENT COAL COMBUSTION**

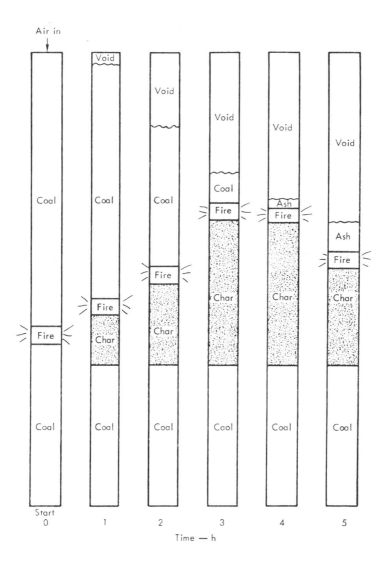

## Char from Subbituminous Coal: Preparation, Characterization, and Rate of Reaction

Char is the solid carbonaceous residue resulting from the heating of coal. In the case of subbituminous coal, the weight of char is 35% of the weight of the raw coal; yet the char retains 70% of the fuel value of the raw coal from which it is prepared. Char is gasified by oxygen and steam in underground coal gasification.

Char prepared from Roland-Seam coal (Wyodak Mine) at 800°C in argon has the formula $CH_{0.14}N_{0.01}$. It also contains 14 wt% ash. As the coal is heated, the surface area first decreases from a value of $\sim$14 m²/g for raw coal to 5 m²/g at 95°C. At 200°C the surface area is 10 m²/g, and at 800°C it is 123 m²/g. The measured rate of reaction of coal char with steam at temperatures from 500° to 675°C is shown in Figure 6.9, where the rate is expressed as a function of temperature (K) by an equation of the Arrhenius form:

$$\text{Rate} = (2.3 \times 10^8)\ e^{-55,000/RT}$$

## FIGURE 6.9: RATE OF REACTION OF CHAR WITH STEAM

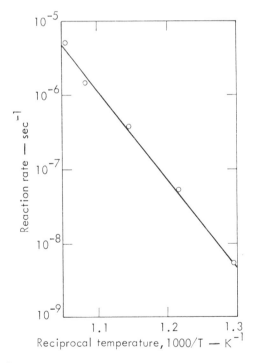

Source: UCRL-50026-75-1

During experiments at fixed temperatures the reaction rate was nearly constant—independent of the fraction of the char consumed, up to 60% consumption. The measured volume of noncondensable gas produced by the steam-char reaction is related to the fraction of char consumed by the measured chemical composition of the gas. At the temperatures of these experiments the gas was essentially two thirds hydrogen and one third carbon dioxide.

These measurements were made on char prepared at 800°C. In actual underground gasification, the char is made in flowing steam and other gases (except oxygen). Furthermore, gasification of the char with steam will start as soon as the char is hot enough.

### Preparation of Activated Carbon from Subbituminous Coal

The fact that the rate of reaction of coal char with steam did not decrease even after 60% of the carbon was gasified could be explained by a char structure which maintained a constant surface area. The surface area was measured as a function of char "burnup" as shown in Figure 6.10 and it

**FIGURE 6.10: SURFACE AREA vs CHAR BURNUP**

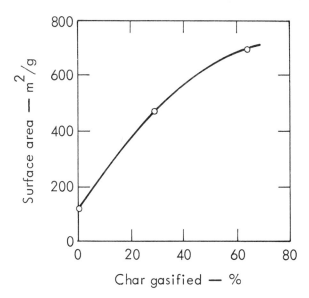

The gasification of char with steam increases the surface area of the char. The gasification was at 675°C, and the surface area was measured by the B.E.T. method using nitrogen.

Source: UCRL-50026-75-3

was found that the surface area, as measured by nitrogen adsorption, actually increased more rapidly than the constant-surface-area model would have predicted. When 65% of the char was gasified, the surface area reached 700 $m^2/g$.

This is a fortunate result for underground gasification, since high surface area means high reactivity, and "carbons" with high surface area can remove contaminants from gases and liquids.

## MATERIALS RESEARCH

### General Metallurgical Considerations

Both the technical and exonomic aspects of minimizing high-temperature effects by using high-temperature alloys, extra heavy sections of conventional casing, and water cooling have been considered. Extensive substitution of conventional materials (low alloy or plain carbon steels) with high-temperature corrosion-resistant materials is economically prohibitive. In addition to high temperature corrosion problems, exposure to elevated temperatures, especially temperature cycles, will undoubtedly cause separation of the encasement cement from the casing possibly causing serious leakage of gas. Thus, the pipes will have to be cooled.

On an engineering and cost analysis basis, water cooling of the production gases is feasible. There are two possibilities: direct cooling of the production gases by water spraying or direct cooling of the pipes by having two concentric casings serve as a water jacket. For the Hoe Creek gasification experiment, water spraying of the production gases is planned. To cool 1 mol of gas by 100°C will require $\sim 0.1$ mol of water. Although, of the two systems, the water-spray system is technologically the simplest and probably the most expedient one to use, it may lead to severer aqueous corrosion.

Although much information exists on the behavior of high temperature, corrosion-resistant alloys in oxygen-rich environments, there is very little information on the behavior of these materials in carbonaceous atmospheres. In coal-conversion systems there may be a mixture of corroding species and the concentration of these may alternate between oxidizing and reducing (relative to some given alloy surface). Resistance of these alloys to corrosion is due to the protection offered by their naturally formed surface oxide films, and these films should deteriorate under reducing conditions.

The metallurgical changes resulting from exposure of eight commercial alloys in high temperature/high pressure, carbonaceous gas atmospheres were evaluated. Either pure carbon monoxide, or pure methane, or a carbon monoxide/methane mixture was charged into the exposure pressure vessel. As the temperature was raised, these decomposed to carbon monoxide/carbon dioxide, methane/hydrogen, and carbon monoxide/carbon dioxide/methane/hydrogen mixtures, respectively. Exposed samples were examined, mainly by light and scanning-electron microscopy.

Evidence of corrosion was apparent in all exposed samples. A continuous grey surface layer covered a granular light-etching sublayer. In many samples precipitation and/or solute depletion occurred below these first two layers. Preoxidation affected the extent of microstructural degradation, and depending on the preoxidation condition could be either beneficial or detrimental. Surface finish was also significant.

Samples exposed to actual coal-combustion products were also examined. Samples were exposed in 305-mm combustion tubes for 90 min.

Both casing-type steels and high temperature alloys were screened. The room-temperature yield strength dropped from 758 to about 400 MPa as a result of the single, short-time temperature excursion. Changes also occurred on exposure of 310 stainless steel ($\sim 25$ wt% Cr) to a maximum temperature of about 1100°C. Film formation and subsurface precipitation were also seen. It is clear that exposure to excessive temperatures will cause loss in structural integrity of in situ pipes even for the more "exotic" materials.

Possibly the use of special cements and techniques will minimize the tendency for separation of the cement from the casing. Coatings, e.g., stainless cladding on mild steel, are also possible as a means of cost reduction for elevated temperature environments.

### Pipe Survival Stress

Analysis of the response of structural pipe to formation loading and displacement during subsidence indicated the following postulated modes of failure:

1. Combined bending and tension.
2. Collapse under external uniform pressure and axial tension.
3. Local buckling of the pipe wall due to bending.

Of the above three modes of likely structural response, only combined bending and tensile effects were found to be of practical interest. The conservative component of bending stress was assumed to be caused by a horizontal displacement of the formation, allowing zero slip between the soil and the pipe. The tensile-stress component was derived from the effective pipe elongation due to the subsidence frictional drag.

## SPECIAL DESIGNS

### Combustion Schemes

There are three basic ways in which the gasification agents can be contacted with coal during underground gasification. The contacting can be such that the "flame front"—the hottest area where the gasification reactions occur rapidly—advances in the same direction as the gas flow; e.g., forward burning. The contacting can be such that the flame front advances in a direction opposite to the direction of the gas flow; e.g., backward burning. Finally, the contacting can be such that the flame front advances perpendicular to the direction of the gas flow, as in the stream method of contacting or the borehole producer method.

The particular gasification scheme may also require special techniques. For example, consider the gasification schemes illustrated in Figure 6.11.

## FIGURE 6.11: COMBUSTION SCHEMES

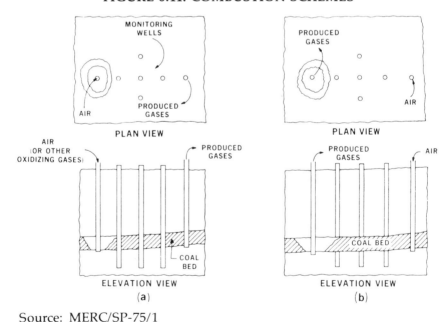

Source: MERC/SP-75/1

Forward combustion is illustrated in Figure 6.11a and Figure 6.11b illustrates reverse combustion. In general, all the interdependent variables such as fluid flow, stress and temperature distributions are different for each scheme. The longwall generator concept illustrated in Figure 6.12 is being investigated in the field by MERC. It is apparent that each gasification scheme and set of field conditions require special designs.

## FIGURE 6.12: LONGWALL GENERATOR CONCEPT

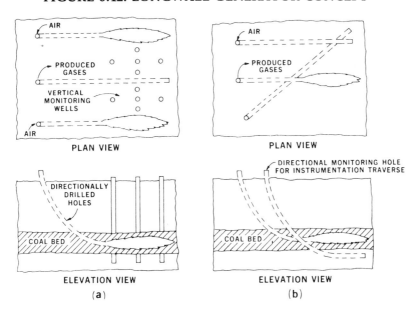

Source: MERC/SP-75/1

### Additional Combustor Designs

Figure 6.13 depicts a possible design which might be suitable for a 100 megawatt in situ combustor. The design, which resembles a retreat longwall mining operation, would require the preparation of underground headings and face area (depicted here as triple entries to satisfy conventional mine engineering requirements) in a horizontal coal seam. The entries circumscribe a coal block (3,000 by 4,000 feet) which is to be completely burned. The outer entry in the headings and face area is filled with fly ash to thermally isolate the combustor from the rest of the coal seam. The two outer headings serve as intake air courses through which air is fed to the burning face. The center heading is a return air course into which combustion products are channelled and then brought to the surface through one or more boreholes cased with high-temperature insulating material. A surface manifold connecting the borehole leads directly to the powerplant, which is situated adjacent to the combustor in an area protected against possible ground subsidence. The powerplant (not depicted in the figure) would have facilities for generation of high-pressure steam and electricity, and control of air pollution, including removal and recovery of particulate matter from the exhaust gases.

## FIGURE 6.13: LONGWALL DESIGN FOR A 100 MEGAWATT IN SITU COAL COMBUSTOR

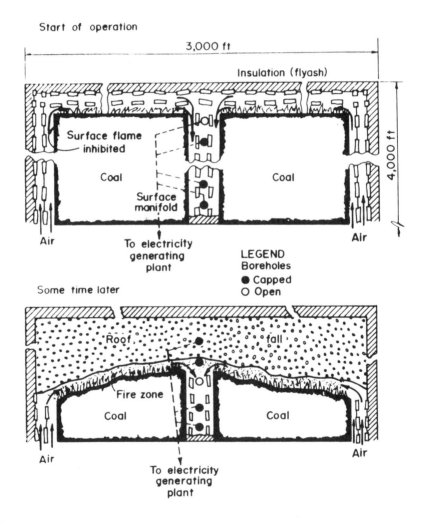

Source: PB 241 892

The cooling effects on the outer headings by the intake airflow, combined with the use of a suitable flame-inhibiting surface coating on the coal block, will help to confine the coal burning to the face area. However, other roof control methods may have to be developed to insure the long-time integrity of the air course. As the burning face retreats, roof falls (either due to natural causes or stimulated by explosive charges) will maintain the airflow against the burning face. Remote fly ash filling techniques through other surface boreholes (not shown) could also be used to help direct the airflow to the fire zone (7). Alternatively, these and other boreholes could be used as auxiliary air intakes to assure a uniform supply of air across the combustion face.

This same overall design with suitable modifications might even be applicable to steep sloping coal seams. In this case, vertical shafts would replace the intake headings and exhaust boreholes.

Another design possibly suitable for hilltop coal seams is shown in Figure 6.14. Here a multiple entry heading is driven straight through the coal

**FIGURE 6.14: HILLTOP DESIGN FOR 100 MEGAWATT IN SITU COAL COMBUSTOR**

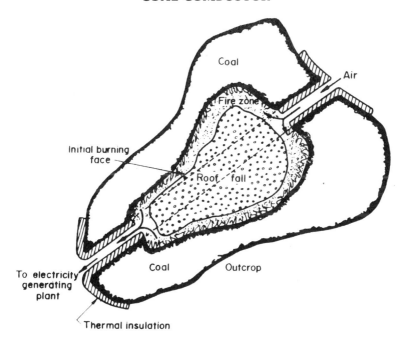

Source: PB 241 892

seam between two points at the outcropping. Fly ash in the outer entries in the vicinity of the outcroppings outlines a burning face which regresses outward from the initial entry in a semiplanar fashion. As the volume of coal consumed increases, roof fall and remote fly ash filling techniques can be employed to maintain an adequate air supply to the burning coal surface.

### Coal-Igniter Design

A 28-kW Petrotherm oil filter heater (Figure 6.15) is to be used to supply hot air for coal ignition in the Hoe Creek gasification experiments. The heater shown in Figure 6.15 was modified by surrounding the three heat-

**FIGURE 6.15: PETROTHERM FIELD HEATER**

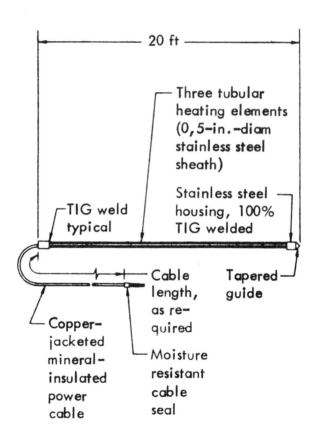

Source: UCRL-50026-75-4

ing elements with a stainless steel tube and wrapping this tube with a 12.7-mm (0.5-in.) thick refractory fiber blanket (Figure 6.16). The air flow passes inside of the tube. This tube acts both to confine the air flow and to contain the radiation heat flux from the heating elements. Confining the air flow increases the convective heat-transfer coefficient, thus decreasing the heater-to-air temperature difference. For the stainless steel tube to effectively contain the radiation from the heater, it must be insulated. A calculation was made for a refractory fiber insulation blanket with a thermal conductivity of 0.069 W/m·°C (0.04 Btu/h·ft). This insulation is sufficient to maintain the tube temperature within 20°C (36°F) of the heating-element temperature. Hence, the effective convective heat transfer area is increased by 233%, further decreasing the heat to air temperature difference.

**FIGURE 6.16: CROSS SECTION OF THE FIELD HEATER**

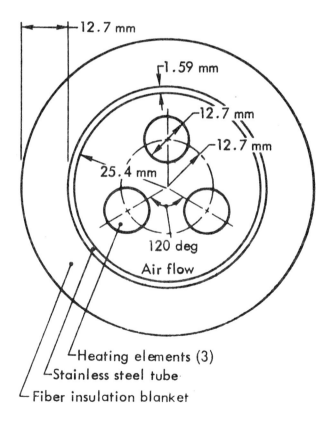

Source: UCRL-50026-75-4

At full power, this igniter design will heat a 0.0189 std m³/s (40 scfm, 1 atm, 60°F) flow from 10°C (50°F) to 951°C (1744°F) in the 10°C (50°F) in situ environment. The corresponding maximum heating element temperature is 1132°C (2070°F), and the corresponding pressure drop is 2.8 kPa (0.41 psi). The calculated steady-state temperature profiles are shown in Figure 6.17. Heat losses to the environment are 17% of the heater output.

**FIGURE 6.17: STEADY-STATE TEMPERATURE PROFILES FOR THE IGNITER**

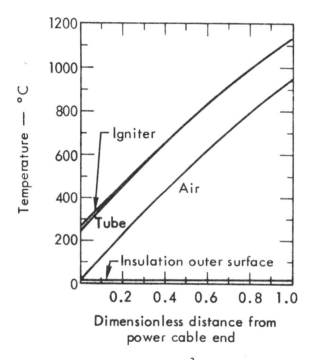

Air flow = 0.0189 std m³/s (40 scfm), air inlet temperature = 10°C, and surrounding temperature = 10°C.

Source: UCRL-50026-75-4

**Flare Design: Thermal, and Flow Analysis**

The internal flare stack for flaring the production gases is shown in Figure 6.18. Design calculations are needed to determine if:

1. The heat losses are high and combustion cannot be sustained.

2. High stack temperatures will cause material problems in the refractory hanger.
3. The induced air flow is insufficient to insure the complete combustion of the production gas.

**FIGURE 6.18: SCHEMATIC OF THE FLARE STACK**

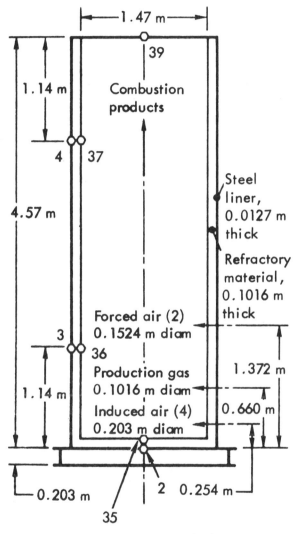

O Indicates important node locations, numbers are node numbers

Source: UCRL-50026-75-4

## SPECIALIZED TECHNIQUES

### Use of Lasers in Drilling

Recently, a number of different types of lasers have become industrially useful tools. There appear to be two main fields where a laser could help in gasifying coal underground. One is in surveying, where a small helium-neon laser could position boreholes, could measure verticality and straightness of holes, and could guide special tools to drill offset holes. Such a laser has an advantage over a theodolite or a level, in that, instead of viewing along a line, its extremely narrow beam of light is easily visible over hundreds of feet. To position some device with respect to this straight line, one needs only to observe the beam on a white target (or one can employ an automatic photocell device). A trained operator is not required.

The other could be in the linking of boreholes, an application for a high-power carbon dioxide laser. Except for drilling, conventional linking procedures do not produce straight line paths between the boreholes to be linked. Carbon dioxide lasers, either alone or, more likely, in conjunction with electro-linking, might be useful to control path geometry.

Tunneling by lasers has been studied but drilling of deep holes in coal seams apparently has not. The power output of industrially applicable systems, usually around 1 kilowatt, is expected to reach 10 kilowatts soon. Even such power levels, however, would still not be high enough to vaporize sufficient amounts of coal—i.e., to drill a hole in coal in the same fashion as lasers are now used to drill microscopic holes—at acceptable drilling rates. Although such an approach would be straightforward, the method would most likely encounter problems in very deep holes, because the vapor will eventually block the beam from the laser either by scattering or by absorption. Continous drilling action will be impossible without some provision for purging.

Alternatively, one could supply air or oxygen to the coal face using the laser only to initiate and to direct the combustion process—i.e., a combustion-drilling process. The problem of scattering by smoke and of absorption by the carbon dioxide in the combustion products would remain to be solved. One could, for example, think of a "drill" pipe with a laser beam going down its center and air being blown through it to the face. The drill pipe would carry a reamer rather than a bit and the combustion would take place a few feet ahead of the pipe. The hot gases would then return to the surface in the reamed space around the drill pipe. Obviously, a severe problem will be caused by the hot combustion gases, which will heat up the drill pipe and probably cause the coal around the pipe to continue to burn. The problem perhaps might be solvable by adding nitrogen to control the oxygen content of the air.

A simpler solution could be to direct the laser at the coal face in an oxidizing atmosphere and to burn until absorption blocks the beam. Then one would flush with nitrogen and repeat the process. Experiments would have to show how deep a hole could be drilled by such a method. An intermittent process such as this would be well suited to lasers that will emit very-high continuous-wave power, but cannot sustain their full output for more than a few seconds or minutes. Certain gas-dynamic carbon dioxide lasers—at present mostly considered for weapons—are of this class.

### Linking Methods

Electrolinking: The process of electrolinking, the electrocarbonization of coal, was first tried in the United States in 1947 in the laboratory of the University of Missouri; field tests were undertaken subsequently under a cooperative contract between the University of Missouri, and the Sinclair Coal Company of Kansas City. More recently, in 1951 at Gorgas, the Bureau of Mines, in cooperation with the Alabama Power Company, further investigated the electrolinking and electrocarbonization of coal underground. In England, about 1952, study and investigation were also directed to the application of the technique to underground gasification processes. In the Russian work, which began prior to World War II, electrolinking-carbonization was successfully applied in large installations.

In the process of electrolinking, electrodes are installed within the coal bed at a given spacing. Passing an electric current between the electrodes carbonizes the coal to form a path of increased permeability through which gases may be pumped. An unlined hole was used and the electrode and electrode stem, which together were approximately 150 feet long, were kept under tension by counterweights to prevent buckling and striking the side of the hole. The counterweight was about 80% of the gross weight of the electrode. In this installation, electrode spacings of approximately 150 feet were operated successfully.

The initial resistance in each test was approximately 14 ohms, or approximately 0.1 ohm per linear foot between the electrodes. The resistance decreased gradually to 3.5 to 4.0 ohms, and then decreased suddenly to approximately one ohm, after which a relatively steady state was achieved. The initial rapid decrease from the 14-ohm level resulted primarily from a rapid decrease in contact resistance between the electrodes and the coal bed, while the period of slower decrease to the 3.5-4 ohm level should be considered to be the electrolinking period: i.e., the period when a low-resistance electrical path is being established between the two electrodes. The very sharp decrease in resistance from 3-4 ohm level value to an approximately 1 ohm level is attributable to electrical breakthrough, i.e., the establishment of path continuity between electrodes.

The final very slow decrease in resistance with time is considered to be the electrocarbonization period: i.e., the period during which heat generated by current flow continues to carbonize the coal in the seam and to increase the cross- sectional area of the path. The gases evolved during this period are typical of the usual products of coal carbonization, and their analyses change as temperatures within the path increase. Simultaneously with gas evolution, porosity is introduced and the permeability of the path is increased. The periods of time required to attain electrical breakthrough in the three tests were quite different because the current paths were erratic and depended on the character of the coal between the particular pair of boreholes used for each test.

Figure 6.19 shows a pattern of electrolinked carbonization paths among a number of boreholes, and illustrates one difficulty—i.e., the lack of control when employing this technique. The paths shown in Figure 6.19 were determined by test-drilling the site after gasification operations in the electrolinked paths had been completed. The path from BH-8 to BH-12 was electrolinked and gasified first. The path from BH-14 to BH-15 was electrolinked and gasified second. The actual reaction path was deflected to the left and BH-15 never actively entered the reaction zone. A path from BH-16 to BH-17 was tried next. Here the actual path formed was from BH-16 to BH-14 at right angles to the path expected.

Thus, the path of the current between two electrodes may be markedly affected by coal bed changes resulting from prior experimental work on the undisturbed coal seam, and the resultant unpredictability of electrolinking in the establishment of single gasification paths appears to offer a major difficulty. But, if multiple paths are to be established simultaneously, as could well be the case in a commercial application of area gasification, the precise path between two individual electrodes may be inconsequential. Estimates made during the three tests indicated that about 80% of the electrical energy had passed through the coal bed during the electrolinking phase, and this proportion rose to about 97% toward the end of the electrocarbonization phase.

Hydraulic Linking: The process of hydraulic linking involves injecting fluids under high pressure into a previously undisturbed coal seam to cause fracturing and consequently an increase in permeability. Hydraulic

FIGURE 6.19: PLAN VIEW OF REACTION PATHS INITIATED BY
ELECTROLINKING AT GORGAS

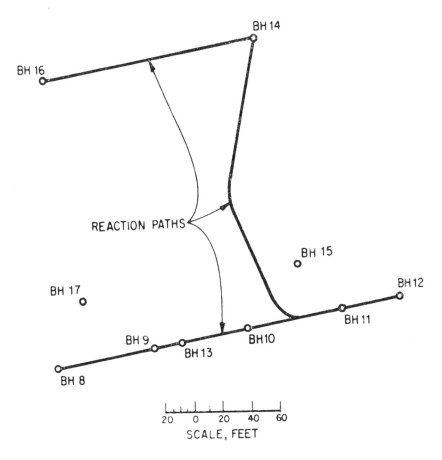

Source: PB 209 274

fracturing techniques have been used in oil-field work for many years and
the method has been tried in the United States and in Russia in connection with underground coal gasification.

In the United States, several tests were carried out at Gorgas, Alabama,
by the Bureau of Mines. In the second and final test in 1957 (8), the coal
bed was fractured by injecting at a rate of 200-600 gpm at 900-1,000 psi
pressure, 11,000 pounds of sand suspended in 33,580 gallons of water
containing a gelling agent for stabilization. The fluid fractured the coal,
and the sand served as a propping agent to hold the cracks and crevices

open after release of the pressure. Fracturing continued for about three hours. The injection took place in a centrally located drill hole.

Pressures measured during the fracturing operation at other boreholes some distance from the injection hole indicated that the fracturing fluid penetrated only to about 150 feet around the point of injection. However, increased air acceptance was noted at test boreholes up to 280 feet from Borehole H. Measurements at existing boreholes 400 feet away indicated that fracture effects may have extended considerably farther than was evident from observations at the test boreholes.

Since air acceptance is a measure of the permeability of the formation, tests were carried out at an inlet pressure of 65 psig. Air was pumped in and the air that flowed out of all other holes was measured while maintaining the inlet pressure. Air acceptance tests were also carried out in another area, 800 feet west of Borehole H, before and after fracturing. The air acceptance before fracturing was 6-13 scfm, and after a first hydraulic fracturing, which used a fracturing fluid of oil and fine sand, the air acceptance was 650 scfm. After a second fracturing attempt, with a fluid and a coarser sand, the acceptance at the same location rose to 1,172 scfm. However, the large increase after the second test was attributed to the large number of open boreholes in the vicinity. The conclusion was drawn that in both cases, there was a 50- to 100-fold increase in permeability at the injection borehole. Air acceptance to other boreholes was increased about five-fold. (These effects should last at least several years.) Subsequently the hydraulically fractured areas were successfully gasified.

Since 1960, a number of discussions of hydraulic fracturing have appeared in the Russian literature. For example, experiments in brown coal seams of the Podmoskovnyi Basin (9) established that hydraulic fracturing in many cases has advantages over electrolinking and pneumatic linking. Other reports (10)(11)(12)(13) discuss hydraulic fracturing and its application with some conflicting statements. Klimentov (12), discussing the use of hydraulic fracturing fluids containing sand to create permanent highly permeable fractures, concludes that the best results are obtained when water without sand is used for fracturing. On the other hand, Belova (13) concludes that the permeability can be improved by hydraulic fracturing and injection of sand.

Pneumatic Linking: A coal bed has a natural permeability. If an inlet and an outlet are provided and air is pumped into the inlet under pressure, some of the air will be recovered at the outlet. The pressure drop and the flow rate are a measure of the permeability.

The natural permeability of coals varies with their ranks. Brown coals (or lignites) have permeabilities that average about 1,000 times more than the permeabilities of bituminous coals. The use of natural permeability of the coal, particularly if it is naturally high, permits linking to occur pneumatically without initial preparation.

In the United States, Russia, and England, attempts have been made to take advantage of the natural permeability of coal seams to develop gasification paths without linking; that is, simply by applying the air under pressure, igniting the coal seam, and then gasifying the coal. One of the most successful and widely reported operations of this type was in Russia, in the Moscow Basin (14).

Work has also been done on the use of high-pressure air to fracture coal formations pneumatically as a separate step (15). Generally, it is possible to achieve initial success this way, but upon release of the pressure, the new fractures tend to close, eliminating the temporarily increased permeability. This closing tendency also occurs in hydraulic fracturing if there is no sand to prop the fractures open after the pressure is released.

The air acceptance of a borehole rises rapidly after unit air pressure equals unit weight of overburden so that the bed can be disrupted and can accept air in a volume sufficient to support rapid travel of the combustion zone during gasification. The air pressure required can be determined by the following formula:

$$P = \frac{DH}{144} + 75$$

where P is the required pressure in psig, D is the average weight (or density) of overburden in pounds per cubic foot, and H is the depth (in feet) of the bed to be linked pneumatically. The additional 5 atmospheres (75 psi) is added to insure breakup of the seam.

Bituminous coal beds, such as those at Lisichansk, are only slightly permeable (0.0015 darcy), and their permeability decreases at increasing depths. At Lisichansk, with 300 to 500 feet of overburden, pneumatic linking, even after increasing pressures as high as 600 psig, was not capable of supplying enough air through the coal beds for adequately rapid linking.

At 300-500 feet, bituminous coal beds may have a permeability some 40 to 60 times greater than surrounding strata. Since the application of air under pressure is most effective in the strata already more permeable, the permeability of the coal bed is preferentially increased and the permeability of overburden is not significantly affected. Thus, permeability can be increased in coal seams irrespective of depth if high enough pressures are used.

But this may not be necessarily so. For example, from 1947 through 1952, generators were being prepared in the USSR at about 250-foot depths, where permeabilities of the bed and the surrounding strata were both relatively high and similar, in the range of 0.2 to 1.8 darcys (15). Hence, linking was not sufficiently selective and consumption of air was high, even at relatively low inlet pressures of 150 to 225 psig. Undoubtedly, losses of air of significant magnitude occurred to the surrounding strata.

Some basic data for the Lisichansk high-pressure pneumatic linkage experiences are as follows:
1. Average pressure of air blast—430 to 515 lb/sq in.
2. Average linking distance—39 ft.
3. Average air consumption—295 to 355 cfm.
4. Rate of linkage—10 to 115 ft/day.
5. Air consumption per foot of linkage channel—43,060 cu ft.
6. Electricity consumption per foot of linkage channel—244 kwhr.

Early experiments with pneumatic linkage in Russia used oxygen-enriched air, but the danger of spontaneous combustion and explosion developed in the inlet hole, and linking had to be carried out finally only with an air blast. Spontaneous combustion slows down the linking process, and is dangerous. Research has shown that spontaneous combustion is less likely to develop with high-velocity blasts than with low velocity because the heat of combustion is dissipated in the blast, and temperature rises are restricted. Dirty boreholes and leakage around casings create conditions leading to spontaneous combustion. Clean well-cased, and cemented holes and air blasts in excess of 90 to 115 cfm have been suggested to insure against its occurrence.

## EXPLOSIVE FRACTURING STUDIES

### Purpose

The basic Lawrence Livermore Laboratory (LLL) scheme for in situ coal gasification involves the use of chemical high-explosives (HE) to fracture coal at depth, thereby increasing its permeability. The region of increased permeability can then be used as a retort through which a flame front is passed, gasifying much of the coal. The design of such a retort requires, as a function of distance from the explosive, a prediction of the increased permeability caused by a given explosive configuration at a given depth.

### Experiment Design

Seam No. 1 in the main pit of the Kemmerer Mine, Kemmerer, Wyoming was available for use in this experiment. This seam, about 26 m thick, is shown in Figure 6.20. To minimize surface effects and reflections from the

### FIGURE 6.20: STEMMING AND EMPLACEMENT OF THE EXPLOSIVE FOR THE KEMMERER EXPERIMENT

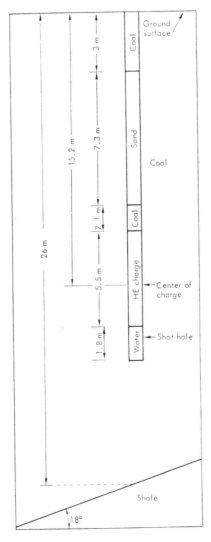

Source: UCRL-51790

shale underburden the charge was located in the center of the seam. On the basis of preliminary calculations a length of 5.5 m and diameter of 10 cm was chosen for the charge. The explosive used was Teledet, which is an ideal explosive (an ideal explosive is one in which the detonation properties are independent of diameter) and was chosen to permit the use of very small-diameter charges in laboratory experiments. The final charge, weighing 59.4 kg, was detonated from one end. Figures 6.20 and 6.21 show the stemming plan and HE configuration.

**FIGURE 6.21: DETAIL OF THE EXPLOSIVE EMPLACEMENT**

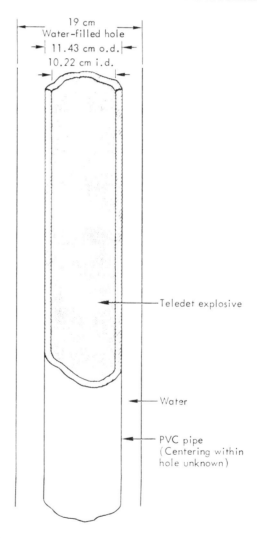

Source: UCRL-51790

Kemmerer subbituminous coal is very highly jointed, with joints (known as cleats) in two sets almost normal to one another. Joint spacing was generally on the order of 1 to 2 cm. Also, the coal was bedded. [A geologic description can be found in Refs. (18) and (19).] However, the bedding appeared to have little effect on the properties of the coal. Compressional wave velocity was between 2.05 and 2.15 km/s according to borehole logs, and 2.3 km/s on laboratory samples, with no effect of bedding observed. Shear velocity was about 1.3 km/s both in the laboratory and the field. Bulk density varied between 1.3 and 1.4 mg/m$^3$ according to logs. The logs are remarkably uniform in depth and from hole to hole, indicating that the bulk properties of the coal are quite uniform. No direct measurements of water content or porosity were made; the coal was water-saturated below about 4 m.

## Calculations

Calculations were performed to investigate whether a sequential or a simultaneous detonation of multiple cylindrical explosive charges, located far from free surfaces, is more effective in fracturing coal. The calculations were performed assuming a 5.1-cm-diameter explosive charge as a source.

In one calculation, the coal adjacent to the source charge was surrounded by a rigid boundary at a radius of 1 m to simulate the effect of other simultaneously detonated charges. In another calculation the coal adjacent to the source charge was surrounded, at a radius of 1 m, by coal with a very low strength to simulate coal already fractured by previously detonated charges. These other charges would then be located a distance of 2 m, twice the radius of increased permeability, from the source charge in the calculation. The failure shear strain, $\epsilon_f$, which is a measure of the cumulative damage sustained by the material, was calculated and plotted vs the radial distance from the charge.

The radius out to where the value of $\epsilon_f$ is greater than or equal to 0.01 appears to correspond to the radius of fracture. The radius vs $\epsilon_f$ is plotted in Figure 6.22 for the cases where the coal adjacent to the source is surrounded by a rigid boundary, fractured coal, or no interface at all. The results indicate a significant increase in fracture radius when the surrounding coal is prefractured and a decrease in fracture radius when the coal is surrounded by a rigid boundary, when compared to the case of no interface at all. Therefore, it is concluded that sequential firing of the explosive charges may be more effective in fracturing the coal.

However, other calculations indicate that for the different detonation sequences used, essentially no differences are noted between sequential and simultaneous detonations. But, using the same parameters, the calculational results indicate fracture enhancement from multiple detonations.

**FIGURE 6.22: FAILURE SHEAR STRAIN vs RADIUS FROM CHARGE**

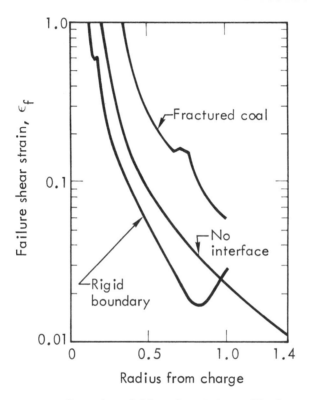

Comparison of failure shear strains resulting from simulated sequential and simultaneous charge detonation.

Source: UCRL-50026-75-2

**Explosives Evaluation**

The evaluation of candidate blasting agents for use in a coal gasification program is essential. Test hardware was constructed to evaluate the compatibility and performance of candidate explosives in the downhole environment. Test conditions representative of the downhole environment are 7MPa (1,000 psi) and 50°C. The compatibility equipment tests a combination of explosive, coal, and downhole water. Detonation velocity is measured to test downhole performance of the explosive.

The compatibility of Teledet, Pourvex Extra, Pourvex EL-836, and DBA 62 T2 was tested in mixtures containing 50% explosive, 25% crushed coal and 25% water. The test conditions were 6.9 MPa (1,000 psig) and 50°C. Drop-hammer-sensitivity and thermal-stability tests were run on the explosives before testing and on the mixtures after one- and two-week exposures to the test conditions. The results of the test were satisfactory. No reactions were observed on the drop hammer, and only trivial changes were noticed in thermal stability patterns.

The ability of two slurry explosives, Pourvex EL-836 and DBA 65 T2, to detonate under pressure was tested. Pressure vessels were constructed from commercial steel pipe, 196.9-mm (7-3/4-in.) inside diameter by 1219.2 mm (48 in.) long. The explosive was packed solidly into the capped cylinder, and a 454-g (1-lb) Pentolite booster placed in direct contact with the explosive at one end. A cover plate was bolted on the flanged end of the cylinder, and the assembly was pressurized to 0.93 MPa (135 psig) with nitrogen from a gas bottle. The pressure is equivalent to 95 m (312 ft) of water. Pins were placed along the longitudinal axis of the cylinder to measure detonation velocity. Test results confirmed that DBA 65 T2 performs essentially the same at test pressure as at atmospheric pressure. Unfortunately, pin signals were lost on the Pourvex EL-836 shot. Observations made on the bunker table, however, indicate a somewhat deeper crater for the Pourvex EL-836 shot.

The performance of Pourvex EL-836 and DBA 65 T2 was measured in 101.6-mm (4-in.) cylinder shots. The energy of Pourvex EL-836 at a volume expansion ratio of 7 to 1 was found to be 20 percent greater than DBA 65 T2.

As a result of these tests and other considerations, Pourvex EL-836 was chosen, because of its energy characteristics and because of economic factors.

### Assessment

The limits of gross effects of the experiment cluster about a radius corresponding to $\epsilon_f \approx 0.01$. Values of $\epsilon_f$ of this order have been associated with microscopic fracture (16). Within this radius one finds:

1. Increased permeability
2. Increased fracturing
3. Decreased sound speed
4. Easier drilling

5. The presence of injected dye for tracing from the shot hole

Also, subtler effects of the experiment have been observed out to a radius corresponding to $\epsilon_f \approx 0.001$. Values of $\epsilon_f$ of this order are close to the limit of calculated shear failure. Out to this radius one finds:

1. Changes in electrical properties
2. Changes in detailed acoustical properties

The correlation may be fortuitous, but it appears that the results of the Kemmerer experiment are consistent with the predictions of Ref. (16), that the explosive did indeed give full yield, and that perhaps one can calculate radii of effect of explosive charges in coal, far from a free surface. Further experiments and calculations, especially on multiple-charge configurations, are required to confirm this conclusion. In addition, the permeability results appear consistent with the predictions of McKee and Hanson (20) and to some extent with those of Stephens (21).

## REFERENCES

(1) P. Averitt, *Coal Resources of the United States*, U.S. Geological Survey, Bulletin No. 1275 (1967).
(2) P. Averitt, *Coal Reserves of the United States—A Progress Report, January 1, 1960*, U.S. Geological Survey, Bulletin No. 1136 (1961).
(3) *Demonstrated Coal Reserve Base of the United States on January 1, 1974*, Mineral Industry Surveys, U.S. Bureau of Mines (1974).
(4) *Strippable Reserves of Bituminous Coal and Lignite in the United States*, U.S. Bureau of Mines, Information Circular 8531 (1971).
(5) *Minerals Yearbook, 1974*.
(6) J. L. Elder, M. H. Fies, H. G. Graham, R. C. Montgomery, L. D. Schmidt, and E. T. Wilkins, *The Second Underground Gasification Experiment at Gorgas, Ala.*, Bur. Mines RI 4808, 72 pp.
(7) L. D. Schmidt, and J. H. Holden. U. S. Pat. 2,710,232, June 7, 1955.
(8) J. P. Capp, K. D. Plants, M. H. Fies, C. D. Pears, and L. L. Hirst, *Underground Gasification of Coal: Second Experiment in Preparing A Path Through A Coal Bed by Hydraulic Fracturing*, Bur. Mines RI 5808 (1961).
(9) G. O. Nusinov, N. Z. Brushtein, M. A. Kulakova, and N. S. Miringof "Experiment in Employing Hydraulic Fracture of a Seam in the Brown Coal Deposits of the Podmoskovnyi Basin," *Tr. Vses. Nauch. Issled. Inst. Podzemn. Gazif. Uglei* No. 5.3-12 (1961).
(10) A. P. Shmarev, and S. N. Yatrov "Group Method of Underground Gasification of Coal Employing Hydraulic Fracturing of the Seam," *Sb. Statei Vses. Zaoch. Politekh. Inst.* No. 26, 69-76 (1961).

(11) I. F. Belova, "Development of an Experimental Battery Generator of Underground Gasification of Coal by the Hydraulic Fracturing of the Coal Seam," *Tr. Tatar, Neft. Nauch. Issled. Inst.* No. 5, 146-52 (1964).

(12) P.P. Klimentov, "Hydraulic Fracture for Underground Gasification of Coal Beds," *Izv. Vyssh. Ucheb. Zaved. Geol. Razved.* No. 10, 97-105 (1964).

(13) I. F. Belova, "Hydraulic Fracturing of Coal Bed Resulting in Formation of Fractures Along 100 M Radii," *Ugol* No. 2, 59-60 (1965).

(14) J. L. Elder, "The Underground Gasification of Coal," pp. 1023-40 (Chap. 21) in *Chemistry of Coal Utilization, Supplementary Volume,* John Wiley and Sons, Inc., New York, N.Y. 1963.

(15) A. Gibb, et al "Underground Gasification of Coal", Pitman and Sons, Ltd., London, 1964, 205 pp.

(16) J. Schatz, *SOC73, A One-Dimensional Wave Propagation Code for Rock Media,* Lawrence Livermore Laboratory, Rept. UCRL-51689 (1974).

(17) D. E. Burton and J. F. Schatz, *Physics of TENSOR74,* Lawrence Livermore Laboratory report in preparation (1975).

(18) W. S. Hunter, Jr., "The Kemmerer Coal Field," *Wyo. Geo. Assoc. Guideb., Annu. Field Conf., 5th, Casper,* 1950, pp. 123-132.

(19) D. H. Townsend, "Economic Report on the Kemmerer Coal Field," *Wyo. Geo. Assoc., Guideb., Annu. Field Conf., 15th, Casper,* 1960, pp. 251-256.

(20) C. R. McKee and M. E. Hanson, *Explosively Created Permeability from Single Charges,* Lawrence Livermore Laboratory, Rept. UCRL-76208 (1974).

(21) D. R. Stephens, *Fracture-Induced Permeability: Present Situation and Prospects for Coal,* Lawrence Livermore Laboratory, Rept. UCID-16593 (1974).

# Technical Problems and Limitations

The material in this chapter was excerpted from PB 209 274, PB 241 892, PB 256 155, UCID-16817, UCRL-51217, UCRL-51686, UCRL-76496. For a complete bibliography, see p 251.

The significant technical problem areas in in situ gasification are discussed in this chapter. The following two observations should be kept in mind if one is to be realistic in the present energy supply environment.

1.  Coal recovery or a maximum recovery of the energy in the coal is important. Certainly one should recover by in situ gasification at least the same amount as that attainable by conventional underground mining, that is, at least 50% which is about the U.S. average for conventional room and pillar systems that leave pillars. Recovery should approach 75-80% (equivalent to longwall or pillar robbing systems), or even better. An in situ process that recovered or utilized only 20-25% of the coal would be unacceptable from the viewpoint of a proper utilization of national resources as well as from the viewpoint of high cost per unit of recoverable energy.

2.  The efficient and successful operation of a pipeline gas plant or a power plant would require a constant supply of a reasonably uniform quality gas feed whether it is from an in situ operation or from a surface plant based on mined coal. It is obviously much easier to achieve this on a well-controlled surface plant using a mined coal than it is for an in situ operation. In situ operations have been subject to problems of quantity and quality of product gas that arise principally from uncertainties involved in roof collapse when the coal is burned away.

Therefore, few problem areas appear to exist that are significant with respect to the generation of electricity from the product gases of underground gasification by the combined cycle, or to the processing of these gases by methanation to pipeline quality. Those that exist are considered to be relatively minor to the problems of in situ gasification. It is anticipated that given a firm supply of product gases from in situ gasification, no technical obstacles remain to block a deliberate program of large-scale commercialization.

## GENERAL PROBLEM AREAS

As stated, the current state of underground coal gasification makes evident the limitations of the technology, most of which are caused by technical problems that have been either difficult to solve or have received inadequate attention. The methods so far developed have been operated on a substantial scale. They can, in fact, produce a combustible gas but not on a continuous basis, or at a constant heating value. Substantial amounts of the fuel value in the coal can be recovered but not at consistently high levels (i.e., 80-90%). Furthermore, none of these methods can be controlled to the extent usually achievable in aboveground processing, so their results are unpredictable. Land subsidence and groundwater contamination problems have not been solved, but this may be attributed to a lack of focus of attention on these problems.

### Combustion Control

Control of the reactions of the coal with the gasifying agent is essential for control of the heating value of the product gases and of the level of recovery of the coal heating value. Ideally, the gas/coal contacting mechanism should be such that the coal in situ is gasified completely, the carbon dioxide produced by the reaction of carbon monoxide with the steam is reconverted to carbon monoxide by reaction with hot carbon, and all free oxygen in the inlet is consumed. Furthermore, the mechanism should be such that contacting efficiency is not affected by roof collapse, if this should occur, and its character is retained even when most of the coal in a panel has been gasified. The problem then is to establish and retain such a contact mechanism.

Actually, gasification runs usually begin with the highest levels of heating value in the gas being produced initially, continue with a gradual loss in heating value as time goes by and as the underground void spaces increase in volume. The trend is a reflection of the increasingly poor contact of gas with the coal face, i.e., too large a combustion volume at the same reactant gas input rate results in inadequate contact between reactants and reduces the reaction rates. A typical set of data is shown graphically in Figure 7.1. Heating value is plotted against time for a single air-gasification run between two boreholes during the operations at Gorgas, Alabama. Also, in the same test and in many others it was noted that maximum coal utilization occurred at the oxidizing-gas inlet regions, where oxygen could readily react with the first hot carbon contacted.

**FIGURE 7.1:  HEATING VALUE vs TIME FOR AIR GASIFICATION**

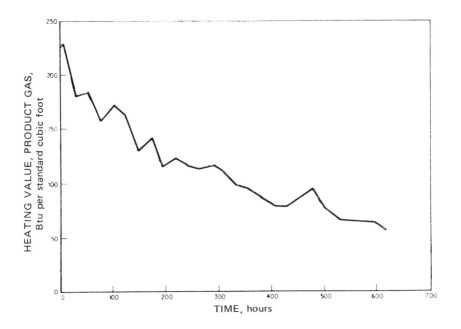

Source:  PB 209 274

Another reason for poor gas quality is the secondary effect of having too large a void combustion volume at the coal face. As the reaction rate decreases, it becomes increasingly difficult to keep face temperatures above the threshold value (900-1000°C) required for effective combustion of coal values. As the combustion chamber enlarges, the inert strata on the roof and floor of the coal seam are uncovered and further increase in void space can take place only along the direction of the coal seam. As the cavity progressively enlarges this way, the roof and floor surface area is increasingly exposed. Through this area heat is dissipated to the strata without a compensating heat release from the combustion of the coal. An increased supply of heat to cope with these increased losses and thereby maintain a sufficiently high temperature at the coal face is required, and this would come about through an increase in the oxygen level in the source. Thus, the ultimate size which the gas-flow channel can reach depends on how long the heat generated in the area can balance the progressively increasing loss to the rock strata. Specifically, this size depends on the ratio of coal surface to strata surface, and this in turn involves the coal seam thickness, and the rate of heat generation, which depends on the rate of air flow.

Roof collapse, if it occurs, can have deleterious effects on gas heating value. Its occurrence causes changes in the character of the void space in the vicinity of the fire front. This is more serious for stream-type processes operating on flat seams than it is for percolation-type processes (regardless of whether forward or backward burning is being employed). In the latter case, the gaseous reactants or product pass through the collapsed rubble and flow problems can only arise, for example, if the collapsed roof material has a low permeability or the character of the collapse can lead to channelling with consequent uneven advancement of the fire face.

The combustion zone temperature should be maintained such that the gasification occurs at temperatures below the fusion point of the ash. If the ash is allowed to fuse, the slag mass can impede coal-gas contact and the decrease in volume can affect roof collapse.

Theoretically, it should be possible to regulate the combustion zone temperature by proportioning in the reactant gases, oxygen and cooling agents such as nitrogen, steam, or carbon dioxide. In practice, however, this is made difficult to achieve by groundwater seepage and reactant gas losses, both of which are aggravated by roof collapse; and by the absence of reliable temperature-measurement techniques which could be used for feedback control for establishing reactant gas compositions. There are no reliable techniques to control leakage in a positive manner.

Therefore, the problem of providing and maintaining efficient contact on a large scale has not been solved satisfactorily yet. The two most serious problem areas are roof collapse and gradual loss in contacting efficiency through enlargement of gas passage cross sections.

## Roof Control

Not very much work has been accomplished in the area of roof control, i.e., the prevention of roof collapse. The procedure generally has been to carry out the gasification process and let the roof collapse as it eventually will when a substantial roof area is left unsupported as a result of the burning away of the coal.

As already noted, uncontrolled roof collapse causes problems not only in the control and successful operation of the gasification process, but also it can result in surface subsidence above the coal deposit, a matter of growing concern especially in the United States.

Subsidence Control: The problem of controlling subsidence or its effects is of particular concern in an in situ combustion process since roof collapse would undoubtedly occur and indeed might be desirable for fire control purposes. This problem imposes constraints on coal bed location and the type of surrounding strata. Some operating control might be accomplished by remote filling of burned-out volumes with sand or fly ash. These same filling procedures might also be used for disposal of solid particulates recovered from the exhaust gases.

In some cases, the gasification operations can affect the roof rock with beneficial results. For example, in the experiments at Gorgas, Alabama, the high temperatures developed by the gasification reactions caused the roof rock to become plastic, to expand, and to settle down on the mine floor directly behind the burning coal face. The fusing and expansion of the roof rock apparently sufficed to fill all the space formerly occupied by the rock and the coal. Careful examination after the test of the immediate roof rock affected this way indicated that it was a grayish, compact, silty clay shale containing important amounts of pyrite, carbonaceous matter and altered mica, generally distributed parallel to the bedding and causing the rock to appear banded.

Work in other locations indicates that the Gorgas experience is probably a special case, since most roof rocks are more often not appreciably affected by the gasification operations and temperatures. When the void space reaches a critical point, roof structures cave as blocks and pieces of various sizes, which fill the voids. A result often can be the generation of large semi-open areas in which gases can wander about and not make close contact with the burning coal face.

A Russian analysis (1) of the relationship of underground gasification with the geologic and hydrogeologic conditions of the coal seam concluded that the effectiveness and the stability of the gasification process is increased by the presence of unstable rocks (e.g., clayey shales) in the roof, that the effectiveness and stability of the gasification process increases with increasing thickness of the coal seam, and that excess moisture, partings in the coal seam, and fractures in the formations reduce the effectiveness of the process and have a detrimental effect on the gasification process.

As mentioned the other aspect of importance in roof control is surface subsidence above the coal bed. This is a problem that has not been considered important in many areas where underground gasification has been undertaken, although surface subsidence has been noticed and measured in some cases. For example, in the Moscow Basin where a 9-foot-thick bed of lignite 160 feet deep was gasified, a substantial surface subsidence was noted. Also, in Great Britain the subsidence was measured over an area where a 2.5-foot-thick coal bed about 240 feet deep was gasified. Here the maximum subsidence was 0.4 feet in the center of the surface depression.

Russian work has recognized the advantages to be gained from an ability to control the collapse of roof strata over voided spaces. Such work has included investigations of both dry (pneumatic) stowing and slurry (hydraulic) filling for supporting the voided spaces. Laboratory work to select materials for pneumatic stowing, done by the USSR Mining Institute included consideration of sandstones and shales in various sizes. This work indicated that these materials could be used for filling.

Studies of hydraulic or wet slurry filling of open spaces and the use of clay slurries for filling spaces in inclined gas generators indicate that this slurry filling work was not productive.

For shaftless underground gasification systems (i.e., where no men work underground) the only access is through boreholes. Although roof control methods here are probably limited to pneumatic stowing or hydraulic slurry filling, these can be considered in various ways using a variety of different materials and methods of application.

Therefore, methods of roof control are available. The fact that their application in underground gasification systems has not been widespread, or has been unsuccessful is probably the result of inadequate attention to its importance rather than of an inability to solve a difficult problem. Roof control should be considered an important factor in further development of any large-scale underground gasification operation for use in the United States. Avoidance of collapse is important to reduce the effect of water infiltration and gas leakage on the control of underground combustion.

## Permeability-Linking-Fracturing

Underground gasification of a coal bed requires that the coal have sufficient permeability to permit the oxidizing gas to flow through the bed without an excessive pressure drop. If the coal is not sufficiently permeable, however, the permeability of the coal can be increased by linking or joining two or more locations in the seam. The methods presently used for linking are: pneumatic linking, hydraulic linking with water or slurries, electro-linking, and directional drilling. During early tests in Russia, explosives were used in attempts to increase permeability; the explosives were set to detonate ahead of the flame front, but the method was not successful.

The permeability of natural coals varies with their rank. Low-rank coals (e.g., brown coals) have sufficient natural permeability so that underground gasification is practical without linking. Bituminous and higher rank coals, however, require linking.

The problem with the usual electro-, hydraulic, and pneumatic linking methods is that they do not increase the permeability of a coal panel uniformly, but form preferred high-permeability paths between the inlet and outlet boreholes. Since such paths usually are not straight, these techniques make it possible for the less-permeable areas of coal to be bypassed during gasification.

The severity of the problem, however, varies with the linking system. For example, in electro-linking followed by immediate gasification the path after linking is already at an elevated temperature—this is an advantage. To illustrate the reverse, the high-permeability path generated between two electro-linked boreholes is not necessarily straight—this is a disadvantage, especially in a panel when multiple electro-links are being established because there is no way of predicting that direct links will be established between desired pairs of boreholes.

The comparative effectiveness of the various methods is difficult to assess, however, because of the conflicting results that have been reported. The Bureau of Mines has shown that the amount of coal that can be gasified from an electro-linked path is much greater than can be obtained from the backward burning of a hydraulically fractured path. Experience reported by Great Britain has shown the reverse to be true.

## Leakage Control

The control of leakage can be very important, since the loss of a substantial amount of the product gas, or the oxidizing gas, can adversely affect the heating value in the product gas or the recovery of the coal heating value. The severity of the leakage problem varies according to the geological characteristics of the site: i.e., the depth of the coal seam, the permeability of the coal bed, the roof and floor strata, and the faults and fracture zones. Although such characteristics can be identified during the planning stage of the underground gasification operation, the prediction of leakage potentials may still be difficult.

In all the underground gasification operations leakage has been a factor. The work at fairly shallow depths has shown that to keep leakage at a tolerable level, gas pressures will probably have to be limited to 50 psig, or less. Higher pressure levels may be possible at greater depths, or in less-permeable formations. In any case, it is quite evident that underground gasification should never be attempted in a severely faulted or fractured area where leakages might be excessive.

The leakage problem can increase and become serious when roof strata collapse and allow gas to escape into porous sedimentary layers above, especially when gasification is attempted in shallow coal seams. On the other hand, if the roof can be prevented from collapsing, and if gas inlet and exit boreholes are properly prepared and sealed, leakage can be kept low and manageable in non-fractured and non-faulted coal seams.

The Mechanics Institute of the Soviet Academy of Sciences concluded that (a) to reduce leakage it is necessary to maintain as low a pressure as possible in the gasified area; (b) the actual gasification reactions must be carried out at as high a rate as possible; (c) for deeply bedded coal seams, leakages of gas are proportional to the square root of the bounding area of the outgassed space, and for shallow-bedded seams, to the first power of the area of the outgassed space; and (d) of the outgassed areas of various shapes, the least leakage occurs in the area which most approximates a circle.

Further examples of leakage problems have been discussed in more recent literature. For example, it was recommended (3) that to reduce leakages in the underground gasification of brown coals, boreholes should be linked by hydraulic fracturing so that lower pressures could be used during the gasification process. Another report (4) indicated that at the Podmoskovnaya station in Russia gas was lost through inlet, outlet, and drainage boreholes, through leakage in the surface lines, and through leakage into the underground surrounding rock structure. The losses in the underground areas were 55% and could not be eliminated; operating at a reduced pressure was recommended. Other reports (5)(6)(7)(8) discuss gas losses and one (7) indicated that losses at one Podzemgaz station reached 30% or more. Losses increased when the static pressure in the underground generators was increased. Also as burnt-out spaces increased, the gas losses increased.

As is evident, the leakage of reactant gases represents a significant problem. The magnitude of leakage depends on the technique employed, the characteristics of the coal deposit, and on its environment.

## Water Control

Sedimentary-type formations associated with coal seams commonly have strata or layers that contain water, often in large amounts. The coal measures themselves typically contain 1-10% moisture in bituminous-type coals and 30-40% moisture in lignites. Conditions vary, of course, and underground coal mines can be quite dry or can have some considerable water inflow from fractures, or faulted zones, that connect the coal seam with aquifers in the formations above the coal.

Up to a certain point, the presence of water in the seam improves the gas quality, because the water-gas reaction can produce more hydrogen in the effluent gas. But this occurs at the expense of heat, since some of it is used to evaporate the liquid water flowing into the reaction zone and the water in the surrounding strata. Above that point, water inflow can extinguish the combustion reactions.

In the category of disastrous possibilities are such things as uncontrollable entry of water into the coal zones. This could occur if there were several major aquifer systems within the coal beds, each at a different pressure. Thus, the coal beds must be carefully selected. Also if the coal were not sufficiently permeable in the case of being shattered with high explosives, then the reactants could not be injected and the whole process would be unfeasible.

As a general rule, it would appear that if underground water conditions permit conventional underground mining, then they will also permit an underground gasification system to be operated. It is easy to install pumps in boreholes to drain aquifers, or keep water under control, as long as the roof can be prevented from collapsing. Thus, the presence of water in the seam probably does not present a serious problem to the gasification process.

Another problem related to water control is the fact that underground gasification can pollute the surrounding groundwater supplies. In work in the USSR (9), two groundwater effects were noted. One was the increase in temperature of the water in the surrounding region arising from heat conduction, from the hot zone, and from the leaking hot gases. The other was an increase in the soluble salts content of the groundwater. In a region where the structure was clays and sands, the soluble salts content increased from about 150 to about 550 grams per liter. In a region of sand-stones and limestones, this increased from around 500 to about 5,000 grams per liter.

In the United States, the potential pollution of groundwater from such a source could be a very serious problem. Measures to prevent or control it would be needed. For example, it might be necessary to pump out af-fected areas, and to treat the water before discarding it into the drainage system. If roof collapse is avoided, the probability of contamination could be significantly reduced.

## Control of Underground Gasification

As in any process, the first step in the control of any underground gasifi-cation operation is to identify the key operating parameters. The next step is to monitor them so that the progress of the gasification is known and anomalies can be detected. Finally, the appropriate corrective actions must be taken to maintain continuous operation at the desired gasifica-tion conditions:

The important operating parameters are:
1. The location and the shape of the fire front,
2. The temperature distribution along the fire front,
3. The detection of roof collapse and the extent and nature of col-lapsed roof debris,
4. The permeability of the coal seam and of possible debris from roof collapse,
5. The detection of reactant and product leakage and of bypassed coal,
6. The inflow of groundwater,
7. The composition of product gas.

The ability to monitor and measure the operating parameters of the un-derground gasification process is in itself not sufficient to achieve suc-cessful underground gasification. The cause and effect relationships must be known so that adequate process control techniques can be developed.

## Pollution Control

As stated previously, it is necessary to prevent pollution of underground water streams and aboveground air by the combustion products. Under-ground water contamination can probably be best forestalled by prevent-ing leakage from the burning area. Atmospheric pollution control would require processing of the effluent gas stream to remove undesirable pro-ducts, such as particulates, sulfur dioxide, and $NO_x$.

**Energy Conversion**

Since hot gas cannot be transported long distances, an in situ combustion process calls for on site conversion of the thermal energy to a transportable form; for example, electricity. While a number of different possibilities exist, operation of a steam boiler-turbine system is probably the most feasible one. The power plant itself can in all likelihood be a separate conventional power package; however, its interface with the in situ combustor could offer signifcant problems in terms of handling combustion product gases at temperatures in the range of 2000°F. These problems include overcoming corrosion, tar, and dust deposition effects, and thermal breakdown of pipes, valves, and insulation.

## FLAME FRONT CONING

Flame front coning or channeling is of critical concern for any in situ process since it can cause bypassing and inefficient resource recovery (10). In this section the problem of flame front coning due to the spatial distribution of access pipes is illustrated. For simplicity, neglect multidimensional flow effects, differences in gas mobility on the coal and coal ash, compressibility effects, and gravity effects. These are severe restrictions. They simplify the calculations enormously, however, and the results still retain and illustrate the essential features of the problem.

The problem being considered is illustrated in Figure 7.2a. In most in situ processes the access pipes look like point sources and sinks in comparison with the dimensions of the bed to be processed. It can be seen that the direct flow line between the inlet and outlet (path a) can be considerably shorter than flow lines that follow the perimeter of the permeable-region (path b).

For flame front propagation through coal, using cocurrent flow, it is reasonable to assume that there is a direct relationship between the mass of oxygen-containing gas that passes through the flame front and the mass of coal that is consumed, that is, the flame-front velocity is directly proportional to the gas velocity.

Consider the two flow lines illustrated in Figure 7.2a with overall lengths a and b. If a flame front is initiated at the source of each line at the same time, the distance the flame front has traversed along path b by the time it reaches the exit point along path a can be expressed mathematicaly.

A number of very approximate calculations can then be made to estimate

## FIGURE 7.2: SPATIAL PROPERTIES OF GAS FLOW

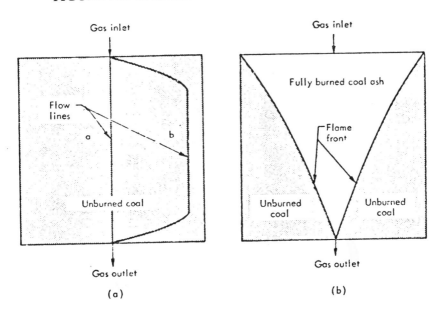

Source: UCRL-76496

the fraction of coal reached at the time the flame front reaches the exit pipe assuming the pressure drop, viscosity, and permeability along the two paths are equal and uniform both in front and behind the burn front. An approximate shape of the burn front at "burn through" can be calculated for the case where the length and diameter of the permeable region are equal (see Figure 7.2b). It can be seen that the shape can be approximated by assuming a V shape for rectangular geometry extending into the plane of the paper, or a cone for cylindrical geometry.

This calculation neglects the effects of diverging and converging flow lines, but still gives a good first understanding of the significance of the phenomenon. If the above "V"-shaped assumption is made and the vertex is defined as being at the exit pipe while the intersection of the legs with the perimeter of the vessel is defined as above a simple estimate can be made of the fraction of coal consumed at "burn-through" as a function of the characteristic dimensions of the permeable region. The results of such an estimate made for the configuration of Figure 7.2 are presented in Figure 7.3. It can be seen that at L/D's in excess of 10, essentially all of the coal is recovered. At L/D's less than 0.1, coal recovery is insignificant. The region of maximum sensitivity, is in the range of L/D = 1.

**FIGURE 7.3: PLOT OF L/D vs F**

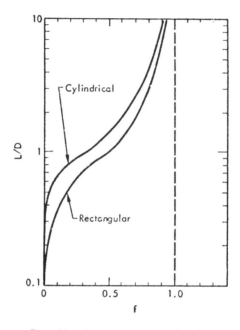

Plot of length-to-diameter ratio (L/D) vs the
fraction of coal burned (f) at the time the
flame front first reaches the outlet.

Source: UCRL-76496

These estimates do not describe the case where gas flow and flame front movement are countercurrent. In the countercurrent case, flame-front-propagation velocities are determined by thermoconductivity parameters rather than the mass balance assumption made here.

It is probable that the permeability of full burned coal ash will be significantly different, and in most cases greater than the coal permeability. This will have a significant effect on flow properties and thus flame front propagation. There are two cases of interest: [1] the coal ash is more permeable than the coal, and [2] the coal ash is less permeable than the coal. In the first case, if one assumes a particular coal bed configuration and a fixed total pressure drop across the bed, the flame front along any particular flow line will continue to accelerate as it traverses the bed. This creates a flame front instability.

As an example, consider the case of totally symmetric and uniform flow, entering at one plane and exhausting at another. One would expect that under ideal conditions a planar flame front would traverse the bed with every point simultaneously reaching the exit plane. This would result in all the coal being burned when the flame front first reaches the exit plane.

Now, however, consider the introduction of a small perturbation, a small region in the flame front preceding the main front. If the coal ash is more permeable than the coal, the perturbation will grow, whereas if the coal ash is less permeable than the coal the perturbation will die out. The first case is defined as an unstable burn flame front, while the second case is a stable flame front. One can readily see that while operating in regions where perturbations grow, less coal will be reacted at the time the flame front first reaches the exit pipe, while the reverse will be true for the case where the perturbations diminish. Therefore, if the coal ash is more permeable than the coal one will get less coal recovery, and if it is less permeable one will get more coal recovery.

## LIQUID PLUGGING

In all in situ operations, the flame front can be propagated underground in two ways which behave very differently in terms of liquid plugging. When the flame front propagates in the same direction as the gas flow (cocurrent), liquid plugging is potentially far more severe than if the flame front propagation and gas flow are in opposite directions (countercurrent). The increased difficulty in the cocurrent case is due to the existence of a condensation front ahead of the flame front which does not occur in the countercurrent case. However, the cocurrent mode of operation is usually preferred because it can be more thermally efficient. Only in the cocurrent mode of operation can one achieve a complete consumption of the carbonaceous material with a single pass of the flame front as well as use the exhaust heat to preheat the part of the formation that has yet to be processed.

Two-phase flow in packed beds with capillary and viscous effects has been studied extensively (11)(12)(13). Liquids can plug the formation by two entirely different physical processes: capillary plugging due to surface tension and viscous plugging due to the liquids having higher viscosities than the gases.

### Capillary Plugging

Partial capillary plugging in a condensation front will not always be a detrimental effect. In one idealized case, the capillary plugging could be maximized along the flow lines where the flame front is progressing most rapidly and it would therefore have a corrective effect by slowing the flame front down in exactly the place where it would otherwise have a tendency to channel. This corrective effect would be operative for "light" capillary plugging, but as the plugging became more severe, it is likely that the plugged zone could either become totally plugged which would cause resources bypassing or could start to have some of the character of a gas pushing a liquid phase. This latter case would also be unstable to gas flow and would cause channeling (1).

The gas pressure gradients and formation crack sizes required to prevent capillary plugging can be estimated. Since explosives can be used to fracture the formation, the formation crack size is potentially an adjustable parameter. The capillary plugging effect depends not only on pore size and surface tension but also on other physical limitations presented in the field, such as overburden pressure and access pipe spacing.

### Viscous Plugging

Viscous plugging occurs when liquids form a condensation front or from general formation seepage from a continuous liquid phase in the pores of the coal seam. Since there is no liquid-gas interface, surface tension effects are not operative. Unlike capillary plugging, in viscous plugging the liquids do not require a threshold pressure drop or flow. They should behave in a manner described approximately by Darcy's law, where any pressure gradient, however small, will cause the liquids to flow. Since it is often desirable in underground processes to make the liquids move, this feature can be advantageously used by designing for the liquids to form a segregated continuous phase (if it is not too detrimental to the rest of the process).

Plugging is a potential problem because liquids generally have much higher viscosities than gases and thus move proportionately slower under the same conditions. This will cause the gases to bypass zones saturated with liquid, which will result in a channeling of the flame front.

The most generally observed plugging of this nature in in situ coal processing occurs when the flame front is propagating horizontally and the liquids accumulate at the bottom of the seam which causes the flame front to channel at the top of the seam. In this case, if one wants to use a cocurrent burn, it is essential that the process be designed so that the liquids can be removed as rapidly as necessary.

## ROOF COLLAPSE

One of the major problem areas associated with in situ coal gasification is the extent of ground movement and its effect on the process. To a first approximation, the area of ground movement during in situ gasification can basically be subdivided into two regions: 1) the region in the immediate vicinity of the gasifier, i.e., within approximately 200-300 ft of the gasification zone; and 2) the area near the surface. The region in the immediate vicinity of the gasification zone affects the extent of gas loss or water influx depending upon the permeability of the strata as well as the degree of caving and fracturing that accompanies void creation. The surface movement affects surface structures, in particular processing equipment, and may also give rise to possible adverse environmental conditions.

### Overburden Collapse

The various theories of strata movement over mined-out areas can basically be divided into two categories (14)(15)(16)(17)(18)(19), arch or dome models, and trough models. The arch theory assumes that as the roof rocks collapse the fractured material forms a stable support in the form of an arch that effectively supports the overburden. The trough theory, on the other hand, assumes the strata deform to fill the mined-out void and that this movement continues to the surface forming what may be called a movement trough. Undoubtedly some combination of both effects occurs with the dominating characteristic (i.e., either trough or dome) being determined by the properties of the confining strata.

### Surface Displacement

The magnitude of the surface deformation associated with subsidence can be predicted. These results are calculated using reported empirical relations for coal mining operations (20).

The depression that develops at the ground surface as a result of underground mining operations is generally characterized by both vertical and horizontal displacement. The vertical displacements generally extend beyond the boundaries of the extraction areas. The surface profile that develops as a result of subsidence (for a given angle of draw) depends on the ratio of the lateral extent of the mined area to the depth of deposit.

Calculations of surface subsidence determined that for a 650-ft-diameter × 100-ft-thick gasification void at a depth of 1,200 ft the maximum vertical and horizontal surface displacement is approximately 60 and 12 ft respectively. These calculations are based on empirical results from other coal mining operations.

The accuracy of measured values of subsidences compared with results from precalculation with profile functions is reported to be good to within $\sim 20\%$ for vertical displacements in the central trough region (20). At distances far from the central trough (where displacement is only several inches) the errors are much larger. The precalculated horizontal displacement in the central trough is generally within $\sim 30\%$ of the observed values.

## REFERENCES

(1) I. V. Korolev, "Interrelation of the Underground Gasification of Coal on the Geologic and Hydrogeologic Conditions at Coal Deposits," *Tr. Vses. Nauch. Issled. Inst. Podzemn. Gazif. Uglei,* No. 8, 51-58 (1962).

(2) J. P. Capp, K. D. Plants, M. H. Fies, C. D. Pears, and L. L. Hist, *Underground Gasification of Coal: Second Experiment in Preparing a Path Through a Coal Bed by Hydraulic Fracturing.* Bur. Mines Rep. Invest. 5808 (1961).

(3) R. N. Pitin, "Some Problems of Reducing the Leakages of Blast and Gas in Underground Gasification of Podmoskovnaya Brown Coals," *Tr. Inst. Goryuch. Iskop., Akad, Nauk. SSSR,* No. 13, 115-24 (1960)

(4) D. K. Semenenko, "Investigations of Blast and Gas Losses, and Methods for Reducing Gas Losses," *Tr. Vses. Nauch. Issled. Inst. Podzemn. Gazif. Uglei,* No. 1, 67-69 (1961).

(5) A. A. Kashkin, D. K. Semenenko, and S. A. Khenkina "Gas Losses at the South Abinsk Station of Podzemgas," *Tr. Vses. Nauch. Issled. Inst. Podzemn. Gazif. Uglei,* No. 8, 12-21 (1962).

(6) I. V. Korolev, "Effect of the Occurrence of Karsts in Coal Deposits on the Underground Gasification of Coal," *Tr. Vses. Nauch. Issled. Inst. Podzemn. Gazif. Uglei,* No. 7, 47-51 (1962).

(7) R. N. Pitin, N. S. Miringof, and V. S. Levanevskii "Influence of Certain Technological Parameters of Process of Underground Gasification of Coals on Magnitude of Gas Losses," *TT-63-11063,* "Underground Processing of Fuels," pp. 97-108 [Translation of *Podzemn. Pererab. Top. Akad. Nauk Tr. Inst. Goryuch. Iskop.13,* (1963).]

(8) R. N. Pitin, "Certain Problems Involved in the Reduction of Losses of Blast and Gas in Underground Gasification of Brown Coal from Moscow Area," *TT-63-11063* ("Underground Processing of Fuels,") pp. 109-18 [Translations of *Podzemn. Pererab. Top. Akad. Nauk Tr. Inst. Goryuch. Iskop. 13,* (1963).]

(9) V. I. Kononov, "Effect of Artificial Heat Source (Subsurface Gasification) on Formation and Composition of Subsurface Waters," *Gidrogeoterm. Usloviya Verkh. Chastei Zemnoi Kory*, pp. 35-51 (1964).

(10) D. W. Gregg, "The Stability of Flame Front Propagation in Porous Media with Special Application to In Situ Processng of Coal, Lawrence Livermore Laboratory, *UCRL-51595* (1974).

(11) A. D. Scheidegger, *The Physics of Flow Through Porous Media* (Macmillan Co., New York, 1960).

(12) T. C. Frick and R. W. Taylor, Eds., *Petroleum Production Handbook*, vol. 2, (McGraw-Hill, New York, 1962).

(13) F. A. L. Dullien and V. K. Batra, *Ind. and Eng. Chem. 62* (10), 25 (1970).

(14) R. S. Lewis and G. B. Clark, *Elements of Mining* (John Wiley and Sons, Inc., New York, 1964) 3rd ed., Ch. 17.

(15) H. G. Denkhaus, *J. S. African Inst. Mining Met. 65*, 310 (1964).

(16) S. D. Woodruff, "Rock Mechanics of Block Caving Operations," *International Symposium on Mining Research*, G. B. Clark, Ed. (Pergamon Press, 1962) vol. 2, p. 509.

(17) R. M. Cox, *Trans AIME 256*, 167 (1974).

(18) G. J. Young, *Elements of Mining* (McGraw-Hill Co., New York, 1946) 4th ed., Ch. 10.

(19) *SME Mining Engineering Handbook*, A. B. Cummins and I. A. Given, Eds. (SME-AIMMPE, New York, 1973) vol. 1, Ch. 13.

(20) G. Brauner, *Subsidence Due to Underground Mining*, Denver Mining Research Center, Denver, Colo., Information Circular #8571 and 8572 (1973).

# Measurement and Instrumentation Systems for In Situ Processing

The material in this chapter was excerpted from MERC/SP-75/1, PB 209 274, SAND-75-0459, SLA-73-0919, UCID-16631, UCID-16640, UCRL-50026-75-1, UCRL-50026-75-2, UCRL-50026-75-3, UCRL-51790, UCRL-51835. For a complete bibliography, see p 251.

## IN SITU INFORMATION REQUIREMENTS

For in situ projects, instrumentation is exposed to harsh thermal, chemical, and mechanical environments. The emplacement process is difficult and expensive and the control center may be hundreds of meters from the instruments themselves. All in all, the instrumentation problems for in situ projects are extensive and require a methodical approach to solutions.

Each gasification scheme and set of field conditions require special designs; however some general information required is common to any process. Prior to considering specific measurements, techniques and instruments, it is important to define the types of information required, the general scope, objectives, and end use of the information. Basic information required for in situ coal gasification process analysis and design is given in Table 8.1. Each of the general categories listed includes many types of data. For instance, stratigraphic data required includes bed thickness, elevations, slope and slope variation, under and overburden characteristics, fracture systems, etc. It is obvious that complete stratigraphic, mineralogic and lithologic analyses will at least require laboratory examination of oriented cores from the zones above, below and through the coal bed. The general directional properties of strata parallel and perpendicular to the bedding planes, including coal beds, influence not only the gasification process but also in situ measurements.

156

## TABLE 8.1: INFORMATION NEEDED FOR
## IN SITU COAL GASIFICATION

1. Stratigraphy, lithology, geology
2. Process input and output parameters
3. Overall process extent
4. Subsurface water configuration
5. Active front location
6. Active front processes
7. Flow distribution (gas and liquid)
8. Materials, structural and flow changes
9. Structural stability of zone
10. Coal elements; mechanical, chemical, physical, and thermal properties

Source: MERC/SP-75/1

The general objectives for in situ measurements and end uses of the data may also vary. Measurements are necessary to answer scientific questions about what really happens during gasification so that realistic physical, chemical and mathematical models can be developed. Hopefully, these research measurements can be greatly reduced once adequate process analysis, design and operation models exist. Certain measurements for process analysis and control will always be needed. Other measurements must always be made for environmental effects due to the gasification process and vice versa. For scientific, process analysis and control, and environmental reasons it is, in general, necessary to make such measurements prior to, during and after the gasification process is complete for a given zone. Such measurements may be absolute values of parameters or relative space-and-time-dependent changes of parameters. Surface measurements alone are generally inadequate for both research and environmental monitoring purposes and measurements are required for subsurface borehole locations.

## GENERAL DIAGNOSTIC MEASUREMENTS

### Monitoring of the Combustion Face

The monitoring process should result in continuous and almost instantaneous knowledge of the position and temperature of the coal face. Methods for achieving this could be based on electrical, seismic, acoustic, and infrared radiometric approaches.

Electrical Methods: No passive electrical methods (i.e., depending on the detection of electrical signals generated by the combustion face) appear to have promise of monitoring the combustion face of an underground gasification installation. Geological prospecting has been done by electric-potential mapping, using the telluric method (ground electric currents), but since the ground currents vary both seasonally and diurnally, they are not reliable. Measurement of the electrochemical potential generated by underground combustion is probably not practical. The maximum voltage to be expected is on the order of a volt and the dipole field will increase as the cube of the distance from the site. Potentials already found in the ground are on the order of tenths of volts, so at relatively short distances from the combustion face the noise background could be inseparable.

Active electrical methods, on the other hand, do show some promise. With an electrothermal approach, one could monitor the position of the combustion face by measuring temperature in the surrounding material. An obvious method would be to use numerous thermocouples strategically placed. This technique would be sensitive to temperature changes. A less sensitive technique would be to use a pair of nichrome wires with fusible shorting links that melt at a known temperature to indicate the approach of the fire zone.

The resistivity of coal deposits and adjacent strata is quite variable, and is largely due to impregnation by ground water, whose resistivity is inversely proportional to its ionic content. (Typical values for resistivity are shown in Table 8.2.) Differences in resistivity might provide a basis for monitoring the combustion face.

### TABLE 8.2: ELECTRICAL RESISTIVITY OF COAL AND ROCK STRATA

| Material | Resistivity (ohm-meters) |
|---|---|
| Coal | 100 to 1,000 |
| Sediments | 10 to 100,000* |
| Wet limestone | 100 to 1,000 |
| Clays | 1 to 100 |
| Graphitic shales | 0.1 to 100 |
| Pyrite ore | 0.01 to 1,000 |

*distribution peaks at 100 ohm-meters

Source: PB 209 274

Differences in resistivity have been used to determine when a drill bit leaves the coal seam. In one case, an electrode was attached to the drill and a tenfold increase in current was measured when the drill entered the more highly conductive strata. In geophysical prospecting, current may be passed between two distant electrodes and an electric potential map of the intervening region obtained. Alternating current is usually employed to obtain greater sensitivity in detection, and to avoid induced polarization. Electrical power on the order of one to ten kilowatts is used. The penetration depth into the ground is approximately one-half the separation between the current electrodes: that is, at the midpoint one-half of the current passes above this depth and one-half passes below.

In a geologically well-surveyed region, a network of current and voltage electrodes could be designed and located to indicate the advance of the combustion zone. The changes in the potential field due to the consumption of the coal or due to dehydration around the fire zone, or both, might generate the required information. Also, carbon and coal resistivity drop rapidly as the temperature is raised. Thus, the hot coal in the fire zone would act as a highly conductive anomaly, and this might be detected either potentiometrically (as described above), or inductively by a large search coil (as in a mine detector).

The use of these active electrical techniques in combination with a high-speed data processor would seem to offer a means of securing instant knowledge of the location and shape of the combustion face, but considerable development appears necessary to reduce any of the techniques to practice.

Active Seismic Methods: Active seismic techniques seem to offer considerable promise in providing a means of locating the combustion front in underground coal gasification because of the drastic changes that should occur in seismic propagation constants at the combustion region. Thus, seismic energy propagating from the unburned region into the burning, or burned, region would encounter a substantial change in elastic parameters. This change would cause some of the impinging seismic energy to be reflected back from the combustion region. This returned energy could then be monitored and processed to yield information concerning the location of the transition region represented by the combustion interface. Extensive work in oil-well and in shallow geological prospecting provides considerable experience and data suitable for application of active seismic techniques to the location of a region representative of the coal/ash interface produced in underground gasification.

In terms of implementing such a scheme, non-explosive seismic stimulus devices would appear to be more attractive than explosive types for many reasons. They would do no damage, present no safety hazards, and their stimulus would be uniform. Two general types of stimulus equipment which would appear to be applicable are the impulse type, such as a drop hammer, and the swept-sine-wave-source type, such as a "vibroseis."

Regardless of which stimulus is used, the seismic source would be installed in one of the boreholes near the periphery of the region that is to be gasified. At two or more locations in the same or other boreholes, elementary seismic receivers would be installed. They would receive "smeared-out" pulses, whose time delay with respect to the transmitted pulse is proportional to the distance the seismic wave travels from the source to the interface and thence to the receiver. For a single receiver, for each such time delay there is a parabolic surface of ambiguity. The position of the combustion interface can be determined by finding the regions of intersection of sets of parabolic surfaces for the various receivers. For a continuous swept-frequency seismic source, the same process can be used: the received frequency becomes a measure of time delay, or travel distance.

Although the system appears to have promise for either type of stimulus, in assessing its utility to solve the combustion face location problem one must consider a number of factors:

1. The seismic wave propagation velocity. The effective propagation velocity must be determined beforehand in order to obtain accurate interface locations.
2. Multipath problems. The receivers will be responsive not only to seismic energy reflected from the combustion interface, but to the seismic energy reflected from all interfaces that the seismic energy encounters as it propagates. Thus, the energy returned from the combustion interface would have to be sorted out from a very large energy return from other interfaces. It may be necessary to preserve the received signal and make comparisons as time elapses. It seems reasonable to expect, however, that the major source of changes in the received signals results from the one thing that significantly changes in the coal medium, and that is the position of the combustion interface.
3. Seismic interference. There will be both man-made and natural seismic disturbances in the medium. The active seismic system must function in the presence of these disturbances which include

microseisms, vehicular traffic, earthquakes, and machinery noises.

The seismic system would probably rely heavily on computer-based data reduction, this reduction being necessary because of the very large amount of data that must be processed to yield quick combustion front position estimates, and also because many geometrical loci must be generated and analyzed for coincidence to determine these position estimates.

Passive Acoustic Methods: The use of a passive acoustic system seems to be a promising approach to locating the combustion front. The basic phenomenon upon which this system relies is the fact that the combustion process releases acoustic energy. Some of this acoustic energy propagates through the gas and some propagates through the coal bed. Whether one thinks of the energy propagated in the solid medium as acoustic or seismic is not relevant.

The attenuation of acoustic energy travelling in a medium like coal at what are normally regarded as acoustic frequencies is well known to be significant. Furthermore the fact that the basic source of the acoustic power (the combustion front) will have components in a wide frequency range means that propagation in the solid medium should certainly be expected for distances of perhaps hundreds of feet. One approach could be to use geophones attached to the walls of the boreholes that are set in at the boundaries of the coal region being gasified to pick up the snaps, pops, and cracks of the combustion.

Consider the electric waveforms produced by the geophones in response to their detection of the acoustic waves in the medium. For discrete, short-duration energy releases at a point in the combustion front, the response at a geophone will occur at a time delay proportional to the distance that the wave travels to reach the given geophone. If the responses of two geophones are compared, there will be a time difference between arrival of the pulses. Knowing the velocity of propagation, it is possible to determine a hyperbolic surface which represents all possible loci for the position of the energy source. In turn, a second hyperbolic surface is generated by the time difference of a second pair of responses. The intersection of the surfaces is a line. Using six pairs of signals from four geophones, one would eliminate ambiguities in position and be able to identify a single point in the medium as the source of the initial release. Repeated application of this process should result in a definition of a surface at which the combustion process is occurring.

If individual events cannot be separated in the geophone signals, the signals from pairs of geophones can be cross-correlated. Such processing of a pair of signals produces a hyperbolic volume of ambiguity and the superposition of the six sets of hyperbolic volumes results in a region of non-ambiguity, which should be the region of combustion.

Although this method appears to have considerable promise, its practical application depends upon several significant influences. Among these are acoustic signals in the medium that are not related to the combustion process, but are due to natural and man-made disturbances. One of the natural sources is the microseisms always present in the earth. Two prime examples of man-made disturbances are the compressors used to provide air flow to the fire, and the noise that air introduces into the medium as it flows to the fire and as it returns from the fire carrying the products of combustion.

Other influences of importance are the properties of the medium, particularly the velocity of propagation of acoustic waves. Because the coal region is bounded within a relatively small volume, the problems of echoes and multi-paths also must be dealt with. Also, the acoustic strength of the source that the combustion process represents needs to be evaluated. If this source proves to be the dominant one in the responses obtained at the geophones, then the technique is promising. On the other hand, if this source is weak, it would render the technique unusable. More likely, the strength could be found to lie in between these two extremes, so that a careful separation of the signals from the natural and man-made background noises would be required.

Infrared Radiometry: Remote infrared radiometry can conveniently be used to detect temperature anomalies of the order of one-tenth of a degree centigrade. It is limited by weather, rain and fog (clouds, if aircraft are used), and by natural temperature gradients. To minimize the latter and to reduce the background as much as possible, predawn observations are often used. The technology is advanced and appropriate radiometers can be obtained commercially. The spectral band from 8 to 13 microns is generally used as it is relatively free from atmospheric interferences and coincides with the maximum in thermal emission from terrestrial surfaces. The technology would be very useful for monitoring of underground gasification of shallow seams. Green, Moxham and Harvey (1) in a study of coal mine fires, report quick detection when the overburden does not exceed 10 meters. At greater depths, direct thermal diffusivity is much too slow and cracks or boreholes would be needed to permit the thermal energy to reach the surface by convection in a reasonable time.

Natural cracks, if present, would be expected to wander, making it impossible to locate the fire precisely.

The only practical way of using this technology for surface monitoring of the gasification process for deep mines would be to sink a number of boreholes into the area. But, in this case, infrared imagery would appear to provide no advantage for deep fire location over the use of thermocouples.

### Monitoring Coal Removal

As in the monitoring of the combustion face, continuous knowledge of coal removal helps ensure that coal is not bypassed.

Gravity Surveys: Gravity surveys have been reported to be one method of studying the amount and extent of removal of coal mass during the burning process. Determination of gravity profiles is routinely done to one mgal accuracy, where 1 mgal = 10 GU (gravity units). Resolution of temporal and spatial changes on 0.1 mgal level would require extremely accurate and competent field work. Therefore, gravity surveys, while conceivable, do not seem very practical for monitoring progress of the underground gasification process.

Carbon Balance: The removal of mass from an underground coal seam can be calculated on the basis of a carbon balance and the coal recovery can be calculated from information regarding the location of the firefront.

Because coal seams are generally of a uniform thickness, a more or less constant quantity of coal is associated with a unit area. This amount of coal can be determined with a fair degree of accuracy during the preliminary development of the gasification system by drill core analysis, chemical analysis, and the like. If the position of the firefront is known at any particular time and the area swept by the firefront is known, the coal recovery can be calculated. Excessive leakage of product gas would affect this calculation, and methods of estimating leakage and composition would have to be developed.

### Detection of Leaks

Leakage of reactant gas can be calculated on the basis of an oxygen balance, a nitrogen balance or preferably by employing a small quantity of an inert tracer gas (e.g., helium).

Product gas leakages from the boreholes and gasification zone through fissures in the surrounding rock and soil are not only wasteful of product gas, but because carbon monoxide is toxic, they represent a health hazard. Detection of abnormally high amounts of either carbon monoxide or hydrogen would probably be the best method of leak detection, since the natural abundance of these two gases in the atmosphere is very low (about 1 ppm). It would be desirable to be able to monitor the gas indicative of leakage at very low levels so that even small leaks could be detected with minimum difficulty. Searching for abnormally high concentrations of carbon monoxide with a device based on infrared spectrometry would appear to be an attractive solution to this problem.

The technique can be implemented in a variety of ways involving either active or passive systems. Spectrometers of various types (nondispersive, conventional and interferometric) have been used to detect trace amounts of various toxic gases. All of the techniques rely on the spectral structure inherent in the vibrational or rotational motions of all molecules, except the noble gases and homopolar diatomic species (e.g., nitrogen and oxygen).

Passive techniques depend on a natural temperature difference between the gas in question and its background. Such temperature differences generally occur for the atmospheric gases, but are somewhat dependent on the local conditions including the meteorology of the area. Active techniques utilize a hot source in the same manner as with conventional spectrometers and, if possible, include a long path through the area in question. Typical detectable concentrations are in the 0.1 to 100 ppm level. Hydrogen cannot be detected by infrared techniques.

Leak detection could be performed by periodically monitoring the area of interest with a portable infrared spectrometer which continually samples the air near the surface of the ground.

Remote detection schemes which would be much less laborious and would permit more frequent, even unattended area surveillance can be envisioned. For example, a centrally located infrared source (probably a laser) could be aimed sequentially at a large number of mirrors (or retroreflectors) placed around the periphery of the area. An infrared spectrometer mounted with the infrared source could pick up the radiation returned from the mirror and detect the presence of carbon monoxide in the path the radiation had traversed. Limited amounts of development have been done on such "searchlight" techniques and further evaluation of their applicability to leakage detection would appear to be warranted.

### Monitoring Product-Stream Components

In monitoring the product stream from an underground coal gasification process, one is primarily interested in the reaction products carbon monoxide, hydrogen, carbon dioxide, perhaps methane and other hydrocarbons and a tracer gas. It is also necessary to monitor unreacted oxygen, water and nitrogen. Each of these gases is normally present at a concentration of greater than 1%, so that a detectability of 0.1% should be generally sufficient.

Historically, high percentages of these gases have been measured gasometrically. A gas sample is taken, passed through a series of reagents, and the volume changes produced on reaction with each reagent are related to the concentration of a given gas. Since such gasometric techniques require a great deal of time-consuming manual labor, the use of automated instrumental techniques in this application seems appropriate.

Three techniques—mass spectrometry, gas chromatography, and infrared spectrometry—are each capable of measuring a number of the specific gases of interest, and have all been used to a limited extent in process monitoring applications. Other, simpler techniques have been developed specifically for process monitoring of some of the gases individually. A number of gas analytical techniques appear to be applicable to monitoring the coal gasification product stream for quick determination of gas compositions. The basic operating principles are reviewed here.

Mass Spectrometry: In mass spectrometry a sample of the gas is leaked into the evacuated ion source of the spectrometer, where it is bombarded by a stream of electrons. The resulting ions are of different masses. Using spectrometer techniques such as quadrupole mass filtering, or electrostatic and/or magnetic deflection, one can selectively detect the various ions that are produced.

In principle, a single mass spectrometer can measure virtually continuously each of the species of interest in this application. However, there are a number of limitations. Since nitrogen and carbon monoxide both have the same nominal molecular weight of 28, a spectrometer of higher resolution, and hence more complexity, is required to resolve the actual small difference in molecular weight between these two gases. In practice, it is difficult to analyze water with the mass spectrometer because water has a strong tendency to be absorbed on the interior surfaces of the instrument. To measure hydrocarbons in general, unconventional spectrometer operating procedures must be employed.

Since the spectrometer operates at a vacuum, a good deal of peripheral vacuum equipment is required. Although low-resolution mass spectrometers are widely employed in industrial leak detectors, few, particularly those with a resolution capable of separating nitrogen and carbon monoxide, are used in process monitoring applications.

Gas Chromatography: In gas chromatography a carrier gas (usually helium) flows through a chromatographic column which is packed with finely divided adsorbent material. After emerging from the column the carrier gas then flows through a detector (usually a thermal-conductivity detector for concentrations greater than 0.1%). The gas sample is introduced into the carrier gas stream and is carried into the chromatograph column. The various component gases are adsorbed onto the column packing to various degrees and, therefore, each migrates down the column at a different rate. The component gases emerge from the column and then pass into the detector separately and in sequence. While in principle, a single chromatograph can analyze a mixture containing all of the gases of interest, the analysis is an intermittent one. The sample is injected and a period of time is required to complete the analysis before the next sample can be introduced. Depending on the detailed design of the chromatographic system, from four to ten analyses per hour of a coal gasification product stream should be possible.

A relatively large number of gas chromatographs are used in process control, but the detailed design of the unit is usually tailored specifically to the type of analysis to be performed. For coal gasification product streams, a number of design modifications would probably be required. A separate column and detector operated with nitrogen carrier gas would be desirable for the analysis of hydrogen. Additional small columns, switching valves, and detectors would be added to the basic unit to permit rapid analyses of water and hydrocarbons.

Infrared Spectrometry: Measurement of gases by infrared spectrometry is accomplished by measuring the attenuation of a beam of infrared radiation as it passes through the sample gas. Molecules with a dipole moment (which excludes oxygen, nitrogen, and hydrogen) absorb infrared radiation at wavelengths that are usually specific for each molecule. Using wavelength discrimination, one can selectively measure one gas in the presence of other gases which also absorb infrared radiation. The so-called non-dispersive infrared spectrometer, virtually the only type used in process applications, achieves its wavelength specificity by using gas-filled cells for filtering. In practice, this means that a separate instrument is need for each gas that is to be monitored. Nevertheless, for the gases for which it is applicable, the non-dispersive infrared spectrometer is a rather rugged and reliable instrument.

Other Measurement Techniques: For the measurement of the oxygen content of process streams, a number of devices have been specifically designed for industrial applications. One is based on the fact that oxygen is a paramagnetic molecule. Another is based on electrochemical measurements of a solution in which the oxygen component of the process stream dissolves and reacts.

Because the thermal conductivity of hydrogen is markedly higher than that of the other gases of interest here, the simple, direct measurement of the thermal conductivity of the process stream should correlate well with the amount of hydrogen present and should be affected only slightly by variations in the concentration of the other constituents.

A number of meters designed to measure the dew point of process streams should be able to measure the water content of this product gas stream. One of the devices involves measuring the heat liberated when the water vapor in the stream is adsorbed onto a highly activiated desiccant. Another controls the temperature of a mirror so that it is always at the dew point of the process stream. Photoelectric sensing is used to detect the appearance of a cloud of moisture on the mirror when the mirror temperature drops below the dew point of the stream.

Commercially available detectors for determining the presence of explosive atmospheres would be usable for measuring hydrocarbons in concentrations greater than about 0.1%. Such explosive gas detectors measure the heat released when the hydrocarbons in the gas stream are oxidized on a catalytic surface. Since the oxidation requires an excess of oxygen, if the oxygen content is low the use of detectors of this type might require the introduction of some air into the stream to be measured. Special precautions would be necessary to avoid erratic response due to poisoning of the catalytic surface if appreciable amounts of sulfur-containing compounds are present in the product stream.

### Sensing Devices for Underground Drilling

To control the direction of the path being drilled, various special instrumentation techniques have been developed, for example:

1. Remote-Reading Compass. A remote-reading device was developed in the United Kingdom in conjunction with the Admiralty Compass Observatory. The device measures the angle between the center line of the borehole and the horizontal component of the earth's magnetic field.

2. Coal-Sensing Device. Another development is an instrument which can reliably detect when the drill bit is approaching the boundaries of the coal seam and which in addition can distinguish whether it is nearing the roof or the floor. Its principle is based on the different degree of scatter of gamma rays in coal and in shale, the usual adjacent stratum.

3. Hole Intersection Detector. The intersection between a vertical and horizontal borehole can be located by measuring the intensity of radiation emanating from a source in the horizontal borehole with a scintillating counter lowered down the vertical borehole. The intensity measurement gives indication of the distance between the vertical borehole and the approaching horizontal borehole.

4. Remote-Reading Inclinometer. Deflection of boreholes from horizontal can be measured by a remote-reading inclinometer, which works on the principle of electrical resistance of a circuit, and utilizes a pellet of mercury to produce a variation in resistance of the circuit depending on the angle of inclination from the horizontal.

### Subsidence and Its Detection

Removal of a mass of coal by combustion is likely to lead to roof collapse, and roof collapse to subsidence in the overlying rock mass up through the ground surface. Subsidence in coal mining districts is well known and preventive measures are usually taken during the mining operations by preserving pillars. Other techniques, such as backfilling with waste or quarry rock and construction of supporting structures are sometimes used. Preventive measures of this kind may not be feasible in the underground combustion operations; therefore, close surveillance of roof collapse would be important for determining the harm that may be done to the gas/coal contact mechanism. Detection of subsidence would be important to indicate that operations to prevent roof collapse have not been successful.

The principal method of determining the amount and extent of subsidence is precise leveling. Repeated surveys of established benchmarks covering the area yield contour maps permitting a graphic representation of the progress of subsidence in the affected area. The main limitations of the leveling method are its slowness and its cost; field crews must periodically survey the area. The detection accuracy of the method is, at best, a few millimeters in elevation, depending on the quality of equipment, experience of the crews and time available for the field work. Thus the leveling method does not detect incipient subsidence nor does it quickly and accurately indicate the rate of subsidence.

It would be desirable to supplement the periodic field surveys with continuous monitoring of subsidence by means of recording instrumentation. However, it would only be economically practical to use such instrumentation to cover some of the sites—they could be selected on the basis of the expected effects or the criticality of their location.

Recording tiltmeters are perhaps the most suitable of the available instrumentation. Tiltmeters of various types (pendulums, water or mercury tubes, for example), are commercially available and achieve resolutions, typically, of a small fraction of a second of an arc, which is far better accuracy than that of the leveling method. The U.S. Geological Survey has used tiltmeters to monitor the inflation and subsidence caused by volcanic activity at the Kilauea volcano in Hawaii.

However, tiltmeters have not yet found general use in monitoring mining operations, but some work in this direction is being done at the Mining Research Laboratories of the Bureau of Mines in Denver, Colorado.

### Testing Plastic Properties of Coal

Coal Swelling: Typically, as coal is heated to 300-400°C, it begins to soften, swell, and agglomerate. At about this same temperature, pyrolysis begins and decomposition gases are generated. These processes continue until about 450-500°C when coal begins to solidify. At about 550°C, resolidification is complete. The solid residue is called 'char" or "semicoke." Upon further heating, the residue contracts until it becomes coke (at about 1200°C), and pyrolysis is completed.

Coal swelling is generally measured by a free-swelling test (2)(3) or a dilatometer (2). The free-swelling test yields little quantitative information, but it is a standard ASTM test and is frequently used. A 1 g ground sample of coal is simply placed in a covered crucible and heated to 820°C. The elevation profile of the resulting coke is compared to a series of standard profiles (3). Each of these profiles is assigned a "free-swelling index" ranging from one to nine. A free-swelling index of one indicates no swelling.

To measure swelling or contraction, one must first define exactly what is wanted. One could measure the size change of a bed of particles, a single particle, a particle excluding the volume in the pores either totally or down to a specified pore diameter or a large block.

Laboratory methods have been devised for these various measurements. Dulhunty and Harrison (4) measured the change in height of a bed of particles in a tube. Mackowsky and Wolff (5)(6) used a microscope to measure the change in particle diameter. Toda (7)(8)(9) measured the specific volume change after cooling and evacuating the coal and then filling the pores at 0.1 MPa (1 atm) with mercury, n-hexane, helium, and methanol (listed in order of accessiblity to smaller pores). Vetterlein and Wong (10)(11) have made measurements using a machined (25 mm diam × 40 mm long) cylinder of subbituminous coal in a dilatometer-type instrument.

The information obtained from a dilatometer is more quantitative than that obtained from a free-swelling test, but it is still somehwat qualitative. Dilatometers measure the volume change of coal as it is heated. Basically, they consist of a cylinder filled with powdered coal and fitted with a moveable piston. As the coal is heated (usually at a constant 2-3°C/min) it expands and drives the piston up. The movement of the piston is then used to determine the volume change of the coal. However, the instrument's design must be considered when interpreting the results.

The Audibert-Arnu dilatometer is designed for use in the coal softening and swelling range of 300-500°C, it uses a pressed pencil briquet of powdered coal that is smaller in diameter than the cylinder. When heated, the coal softens and flows into the annular space. Thus, the initial indication of plasticity is a contraction. This is followed by an expansion as the coal continues to swell. A typical curve is shown in Figure 8.1. The significant

**FIGURE 8.1: PLASTICITY CURVE FOR COAL IN AN
AUDIBERT-ARNU DILATOMETER
HEATING RATE—2°C/min**

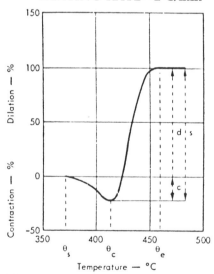

Source: UCRL-51835

points indicated are: $\theta_s$ = softening temperature; $\theta_c$ = initial swelling temperature; $\theta_e$ = final swelling temperature; c = maximum contraction; d = maximum dilatation; s = maximum swelling. Note that the measure of maximum swelling, (s), may not be the true maximum volume change.

In Sheffield and Hoffmann dilatometers, powdered coal fills the cylinder instead of a smaller diameter briquet. However, the diameter of the piston is small, allowing the piston to penetrate the softened coal as it is heated.

It is generally difficult to obtain quantitative swelling information from any of the above instruments. Dilatometers yield quantitative information if the coal does not soften and lose its shape (i.e., for nonswelling or contracting coals).

Coal Viscosity: The viscosity in coal is generally measured with a plastometer. There are several similar instruments, with the constant-torque Gieseler plastometer (2)(12) being the most commonly used. A finely powdered coal is compacted into a cylindrical crucible containing a bladed stirrer. A constant torque is applied to the stirrer as the coal is heated at a constant 2-3°C/min. The angular velocity of the stirrer is recorded in degrees/min or in dial divisions/min (100 dial divisions/revolution). Figure 8.2 shows a typical curve. The significant data points are $\theta_s$ = softening temperature, $\theta_m$ = temperature corresponding to the maximum angular velocity, $\theta_r$ = resolidification point, and $\omega_m$ = maximum angular velocity.

## FIGURE 8.2: PLASTICITY CURVE FOR COAL IN A GIESELER PLASTOMETER

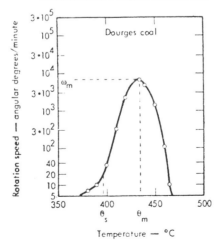

Source: UCRL-51835

The plastometer can be calibrated in terms of viscosity by using glass or other material of known viscosity as a reference (13)(14)(15). This calibration is meaningful if the coal is not too fluid. If the coal becomes too fluid, the stirrer whips the coal mass into a foam, thereby generating erroneously low viscosity values.


## LERC INSTRUMENTATION AND PROCESS CONTROL PROGRAM

### Background

The coal being utilized in the LERC test site is the Hanna #1 seam, a 30-foot thick (9.1m) subbituminous coal lying at the depth of approximately 400 feet (122m). This seam is a major near-surface aquifer. Compressed air was the only gasification agent used.

In the Hanna #1 experiment over a five month period, an average of 1.6 MMCF dry gas/day (45,340 nm³/day) was produced based on an average air injection rate of 1.076 MMCF/day (30,450 nm³day) with an average heating value of 126 Btu/SCF (1,122 Kcal/nm³). Approximately 20 tons (18 metric tons) of dry coal were affected per day to produce this amount of gas. Four times the amount of energy used to run the experiment was produced in the form of combustible gas, liquid organic materials, and sensible heat.

### Experiments for Seismic-Acoustic Measurements

Figure 8.3 shows the approximate location of the burn front during the time period of interest and the location of the three seismic detectors which had been lowered to the middle of the coal seam. The experiment objective was to assess triangulation as a burnfront location technique.

Essentially four different classes of signals were identified: (A) intermittent pulses with different arrival times at different sensor locations, (B) series of pulses with exponentially increasing amplitudes that followed compressor shutdown, (C) continuous steady state noise observed in quiet periods, and (D) background noise clearly identifiable with surface disturbances such as blasting.

Cross-correlation of Class A signals indicated an association with burn area. Failure to record these signals in Well 2 precluded triangulation. Laboratory experiments showed that the signature of Class B signals matched that of cascading water. This implies monitoring of water return

**FIGURE 8.3: WELL PATTERN AND GEOPHONE LOCATIONS**

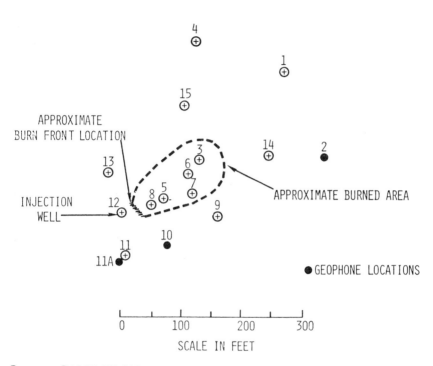

Source: SAND-75-0459

into the burn region during compressor shutdown. Finally, laboratory experiments on burning wood produced a steady state white noise analogous to that obtained in the field (Class C). This laboratory-produced white noise appears to be composed of discrete signals associated with different regions of the combustion front.

While definite association of these three classes of signals with specific activities of the gasification process was not quantitatively established, these experiments did indicate that seismic systems might have application for monitoring in situ processes.

**Surface Resistivity Measurements**

In December 1973, surface resistivity surveys of the Hanna #1 site were made. The direction of the scans through wells 3 and 7 was noted. The Wenner configuration (17) was used for depth soundings and did indicate deviations at the depth of the water table and burn zone, respectively. The Schlumberger configuration (17) was used to obtain an indication of areal resistivity.

Quantitative interpretation of these limited results is not possible; iso-resistivity contours can be inferred. They can be used as an indication of how such a technique might be used to delineate the location and extent of an in situ combustion process.

**Temperature Measurements**

This program also evaluated thermal instrumentation to determine the temperatures and thermal profile associated with a burn front during a forward combustion process (Figure 8.4). A secondary objective was the evaluation of a nonconventional branched multipoint thermocouple circuit (18).

**FIGURE 8.4: THERMOCOUPLE INSTRUMENT WELL**

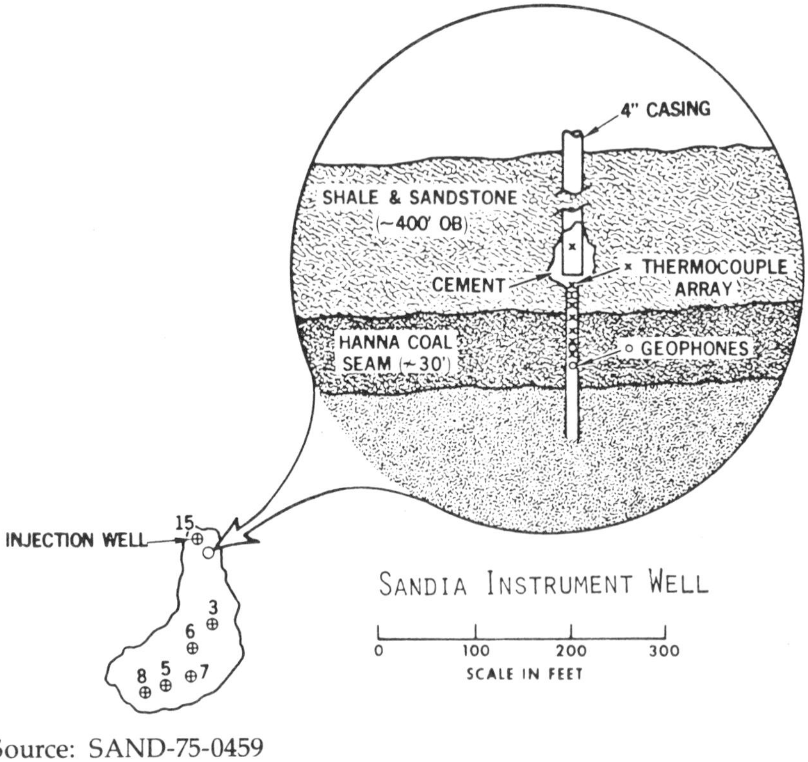

Source: SAND-75-0459

**Hanna #2 Test Plan**

To cover the three phases of the Hanna #2 experiment, a total of 19 instrumentation wells were to be drilled.

In Phase 1, as shown in Figure 8.5, four injection/production and instrumentation wells, respectively, were located. Linkage and burn were originally conducted between Wells 1 and 3, along the path of instrumentation Wells DD and CC (19). The burn front is now directed towards Well 2, so that instrumentation Wells BB and AA are more directly involved. In

**FIGURE 8.5: HANNA #2 EXPERIMENT PHASE 1**

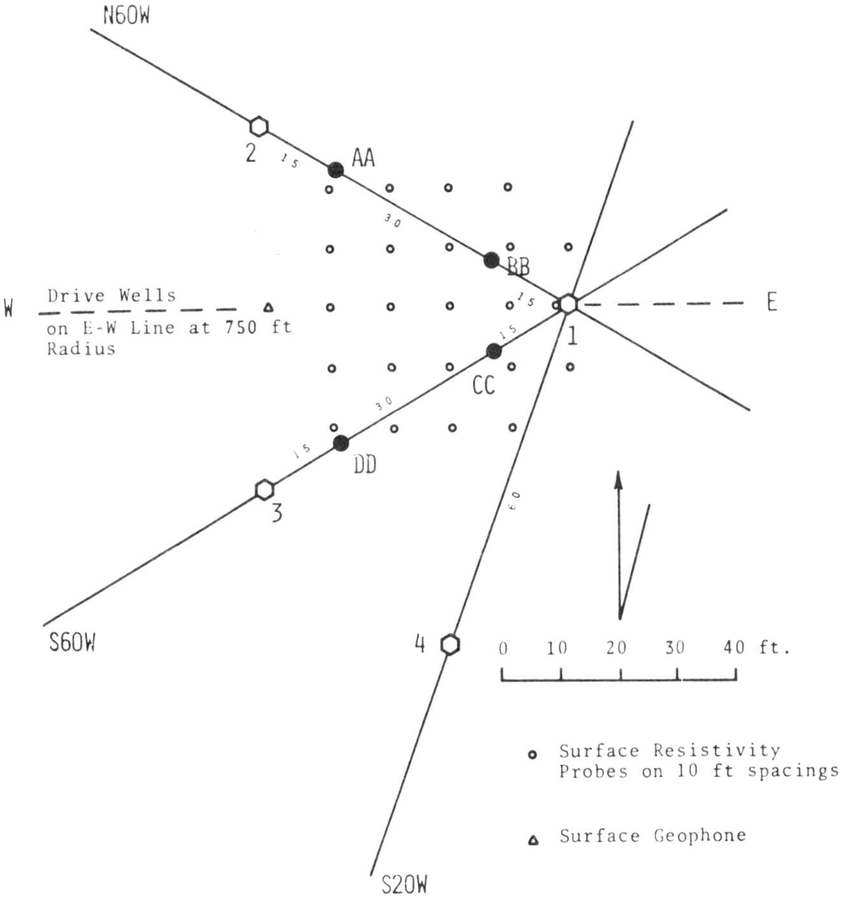

Source: SAND-75-0459

accordance with the initial, limited objectives for LERC's Phase 1 experiment, the instrumentation effort was modest and designed to better evaluate sensor performance and define problem areas to enhance the instrumentation effort on Phases 2 and 3. Even this limited effort produced some very significant results.

The instrumentation well placement for Phases 2 and 3 of the Hanna #2 experiment is much more extensive, as shown in Figure 8.6. The instrumentation well spacing was chosen to monitor the oxygen enrichment as well as the line-drive sweep experiments; both with sufficient accuracy to provide spatial resolutions on the order of a few feet.

**FIGURE 8.6: PROPOSED WELL LOCATIONS, HANNA #2
EXPERIMENT PHASES 2 AND 3**

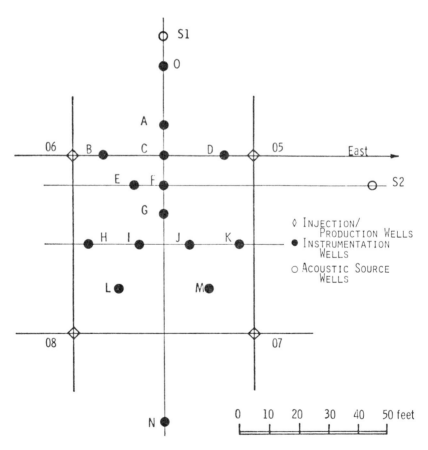

Source: SAND-75-0459

Over 400 sensors are to be included in the surface and sub-surface instrumentation packages. One of the instrumentation wells with its associated sensors is illustrated in Figure 8.7.

## FIGURE 8.7: INSTRUMENTATION WELL

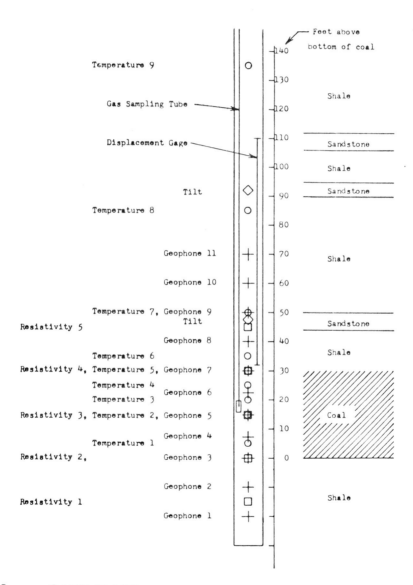

Source: SAND-75-0459

## Hanna #2 Temperature Measurements

The objectives of the temperature measurement part of the instrumentation program are comprehensive. In the 15 instrumentation wells, associated with Phases 2 and 3, about 112 "low", that is chromel/alumel, thermocouples are to be used and 16 "high", or platinum/rhodium, thermocouples. Specially designed branched circuitry is to be used, as shown in Figure 8.8, for these thermocouple arrays. These circuits provide the opportunity for redundant temperature measurements, differential temperatures, and continuous circuit diagnostics.

### FIGURE 8.8: BRANCHED THERMOCOUPLE CIRCUITS

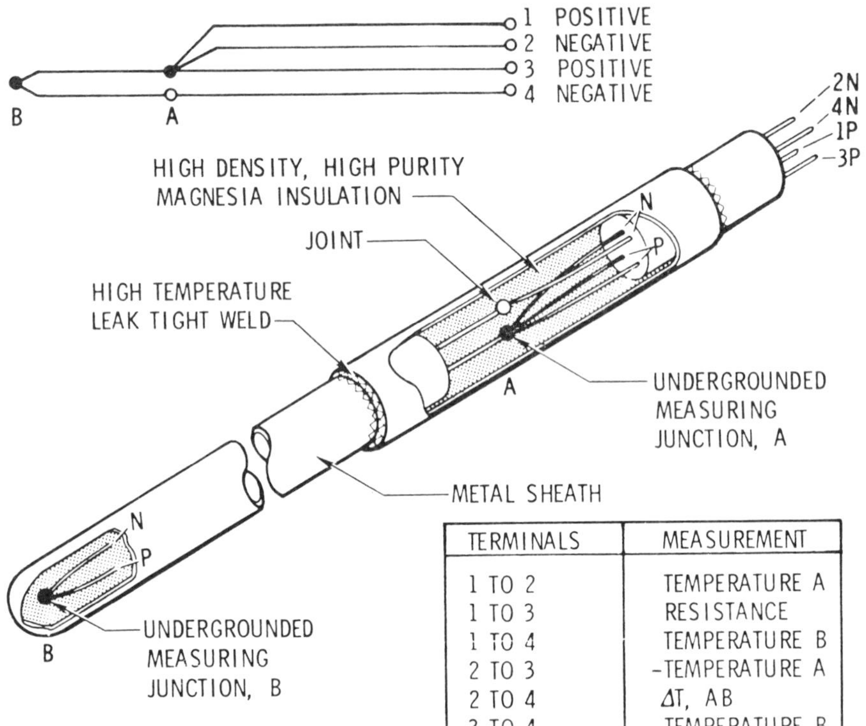

| TERMINALS | MEASUREMENT |
|-----------|-------------|
| 1 TO 2 | TEMPERATURE A |
| 1 TO 3 | RESISTANCE |
| 1 TO 4 | TEMPERATURE B |
| 2 TO 3 | -TEMPERATURE A |
| 2 TO 4 | $\Delta T$, AB |
| 3 TO 4 | TEMPERATURE B |

Source: SAND-75-0459

### Hanna #2 Resistivity Experiments

Surface and subsurface resistivity measurements are used to detect the location of the combustion front. One configuration to be evaluated is shown in Figure 8.9.

## FIGURE 8.9: VOLTAGE MEASUREMENT TO DETECT THE LOCATION OF THE COMBUSTION FRONT

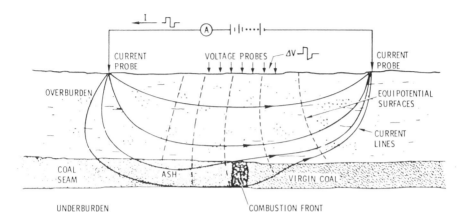

Source: SAND-75-0459

Current for the resistivity measurements is to be introduced into the earth at cased 30-foot wells located approximately 750 feet from each of the sites (Wells 1-4 and Wells 5-8). On Phase 1, one pair of drive wells is to activate an array of 24 surface probes on 10-foot spacings. On Phases 2 and 3 respectively, about 100 surface probes are to be used on 10-foot spacings.

Other configurations to monitor potential changes include subsurface voltage probes in conjunction with the surface current source electrodes and conversely, a one current source electrode implanted in the coal seam with potential changes being monitored at the surface.

### Hanna #2 Seismic-Acoustic Experiments

Geophones are to be located in each instrumentation well, as well as selected surface locations to detect combustion front-associated signals. In these passive acoustic experiments, triangulation techniques are to be utilized to locate source of signal. Emphasis is to be given to associating classes of signals with specific aspects of the gasification process.

Induced seismic-acoustic techniques are also to be evaluated as a means of determining burn front location, as illustrated in Figure 8.10. Initial emphasis is to be on downhole energy sources and geophones. Two-dimensional shadowing models were developed to determine burn front geometry. It is conceivable that this technique, based on attenuation of generated signals as well as knowledge of refracted paths, might ultimately be utilized to assess extent of resources affected by combustion.

**FIGURE 8.10: INDUCED SEISMIC-ACOUSTIC TECHNIQUE**

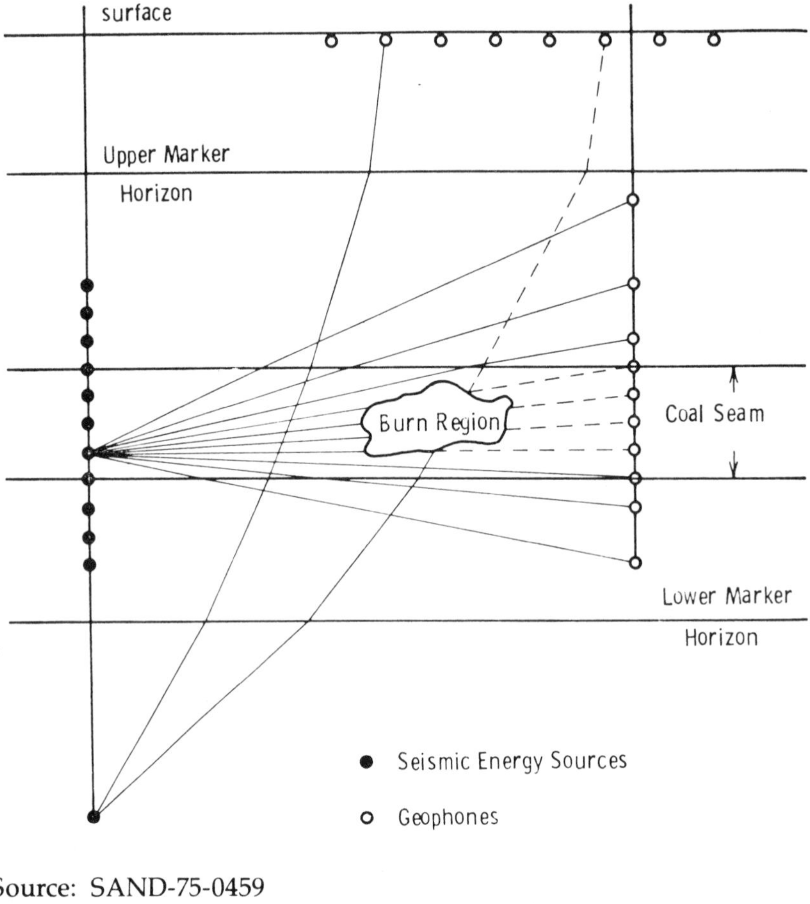

Source: SAND-75-0459

A deep downhole energy source coupled with a limited array of surface geophones is also to be investigated as a means to determine burn front location. The significance of this experiment is that it is the second step for developing a remote monitoring experiment.

**Data Analyses**

The temperature data have proven of greatest value in aiding in the evaluation of Phase 1. Figure 8.11 shows thermocouple data, obtained from Well CC, during the air injection portion of the linkage experiment. Note the profiles through the coal seam, providing an indication of air distribution. The thermocouple system utilized was capable of measuring small temperature rises. The temperature increases observed are thought to be a result of coal oxidation caused by the passing air.

**FIGURE 8.11: REPRESENTATIVE TEMPERATURE PROFILES DURING AIR INJECTION PHASE**

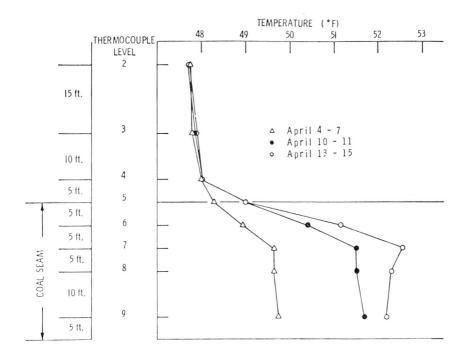

Source: SAND-75-0459

After air injection, as a linkage process was attempted, the coal was ignited in Well 1 and reverse combustion was initiated. Linkage was established via this process but, as the data in Table 8.3 illustrate, a highly localized channel was established. Only Well CC, at the center of the coal seam, measured the location of linkage. Linkage apparently bypassed Well DD. It is possible, however, that Well 3, which was redrilled and whose downhole direction is not established, is not on a straight line with Wells CC, DD and 1 (Figure 8.5).

**TABLE 8.3: TEMPERATURES AT SELECTED LOCATIONS BEFORE AND DURING IGNITION AND REVERSE COMBUSTION PHASES**

| Thermocouple | Location | Observed Range May 15 - June 3 |
|---|---|---|
| BB-8 | Seam center | 51.4 - 52.0°F |
| DD-7 | 5 ft. above seam center | 52.5 - 53.8°F |
| DD-8 | Seam center | 51.3 - 52.5°F |
| CC-7 | 5 ft. above seam center | 54.1 - 54.9°F |
| CC-8 | Seam center | 53.9 - 198.5°F |
| CC-9 | 10 ft. below seam center | 52.2 - 53.2°F |

Source: SAND-75-0459

After linkage was established, the gasification process reverted to forward combustion. At Well DD, closest to the injection well, temperature levels as a function of time at mid-seam and above are presented in Figure 8.12. This instrumentation well did not have provisions for thermo-

**FIGURE 8.12: TEMPERATURES IN WELL DD, PHASE 1**

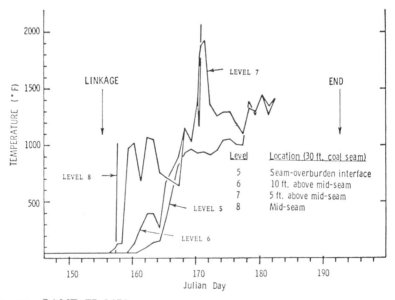

Source: SAND-75-0459

couples below mid-seam. Levels 5-8 indicate a reasonable degree of uniformity in measured temperatures before signal loss occurred, thus implying a desirable areal distribution of the gasification process.

At Well CC, closest to the production well, similar temperature history records are more informative. In this well, one thermocouple location existed 10' below mid-seam. As Figure 8.13 illustrates, the combustion front, when it reached Well CC was at mid-seam and below. As levels 6 and 7 indicate, buoyancy effects ("gravity override," "bypassing") (20)(21) had certainly not overtaken the combustion front lifting it to the coal seam-overburden interface. As this experiment period was ending, the temperature at level 7, five feet above mid-seam, was to rise above the 200°F steam value substantially after level 9, ten feet below mid-seam has reached peak value. Thus, based on these limited data, it appears that the linked vertical well concept of in situ coal gasification may not be adversely affected by buoyancy effects. This should allow a high percentage of the coal seam to be affected by the gasification process.

## FIGURE 8.13:  TEMPERATURES IN WELL CC, PHASE 1

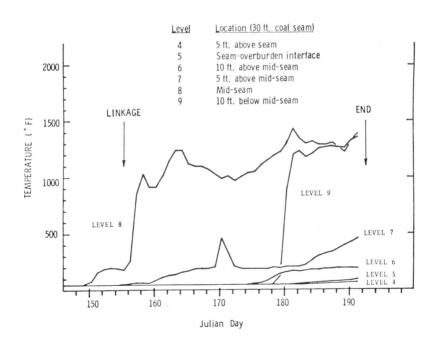

Source: SAND-75-0459

The geophone tapes were examined for signals associated with the linkage and combustion processes. No significant activity was recorded during the air injection or reverse combustion linkage experiments. Discrete signals have been recorded in association with the forward combustion process. These signals are correlatable, but the attenuation through the medium, which is on the order of a tenth of a dB per foot, increases the difficulty of triangulation. A fairly large signal recorded in Well DD, was reduced in magnitude by a factor of five in Well AA, at the same level, some 45' distance. Whether passive acoustic techniques can operate over a sufficiently large distance to be a viable technique is questionable.

Surface resistivity measurements have been made on Phase 1A using surface electrodes and surface voltage probes. The resistivity data have to be correlated with estimates of burn front location and volume affected before the value of this monitoring technique can be assessed.


## LLL INSTRUMENTATION PROGRAMS

Reliable field measurements are essential to the successful development of the in situ process. At the coal-outcrop experiments at Kemmerer, Wyoming new applications of well established geophysical techniques were tested and developed. The following discussion outlines the initial experiments.

### Electromagnetic Methods

Both high-frequency (HF) and low-frequency (LF) electromagnetic (EM) techniques have been used to detect anomalies in regions of geologic interest (22)(23). Therefore, measurements of electrical characteristics could be useful in delineating the extent of explosive fracturing. For electromagnetic methods to be effective, the subsurface regions of interest must differ significantly from the surrounding medium. At Kemmerer, this condition was provided by the large differences in electrical properties of the coal, water, overburden, and underburden. For example, the low-frequency bulk resistivity of undisturbed, water-saturated coal was about 2,000 $\Omega \cdot m$, whereas the water itself had a resistivity of only 10 $\Omega \cdot m$. Therefore, the extent of fracturing should be detectable because of the increased water content (and lowered resistivity) of the fractured coal.

On the basis of these considerations, preshot and postshot EM measurements of the region affected by the HE detonation were conducted (24). Significant changes in EM characteristics were indeed detected and the details of the measurements are presented below. In addition, the feasibility of characterizing the fluid permeability of a region by EM methods was demonstrated. (A subsidiary measurement showed the effectiveness of the EM technique for locating abandoned mine workings.)

## Low Frequency Measurements

Resistivity measurements were made at low frequency (20 Hz) using a dipole-dipole configuration on the surface. Four metal pins were driven into the ground. Two were connected to a current source and two to a narrow-band voltmeter so as to form transmitting and receiving dipoles. Currents and voltages were recorded, both as the dipoles were moved over the earth's surface at a fixed spacing (horizontal profiling) and as the spacing was increased symmetrically about a fixed point. The latter technique is called depth probing and can, in principle, provide an indication of the depth profile of subsurface electrical resistivity.

Results obtained at Kemmerer using the fixed-spacing technique are shown in Figure 8.14. The preshot curve indicates a smooth decrease in apparent resistivity suggesting a homogeneous coal seam with no big cracks or anomalies. The postshot survey, which was done over the same area in the same fashion, produced a curve that shows a definite anomaly covering approximately 6 m on each side of the center line of the shot hole. The break points A and B, are quite distinct.

**FIGURE 8.14: HORIZONTAL PROFILE OF ELECTRICAL RESISTIVITY FOR PRE- and POSTSHOT CONDITIONS**

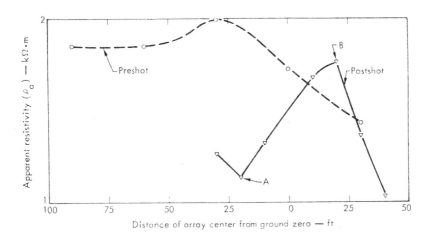

Source: UCRL-50026-75-1

### High Frequency Hole-to-Hole Measurements

A second method of electromagnetic probing involves the use of high-frequency equipment ( $\geqslant 1$ MHz) located below the ground surface. Measurements were made in a set of four equally-spaced holes located on a circle of 4.6-m radius with the shot hole in the center. The holes were drilled to 30 m so the bottoms of the holes were below the coal seam. A transmitting antenna was placed in one hole and the high-frequency EM signal transmitted through the coal, including the shot hole, to a receiving hole on the opposite side of the shot hole.

The transmitting- and receiving-antenna depths were varied to produce a ray-path tracing as shown in Figure 8.15. Holding the magnitude of the

### FIGURE 8.15: TRANSMITTER AND RECEIVER LOCATIONS FOR HF HOLE-TO-HOLE MEASUREMENTS

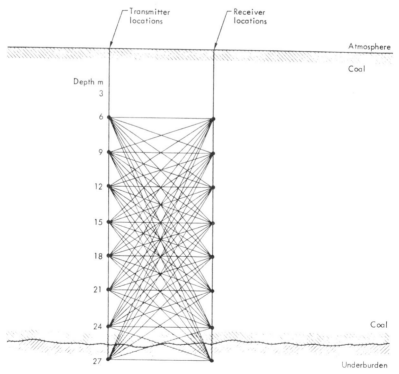

The direct-line paths are shown for all possible transmitter-receiver pairs.

Source: UCRL-51790

transmitted signal constant, magnitude and phase of the received signals were then recorded for each position. The skin depths, or exponential attenuation coefficients (in water-saturated coal), were calculated from the magnitude-data of the received signals. The dielectric constants (for the coal medium) were calculated from the phase shifts, for a 1 MHz change in frequency.

The data were computer-analyzed with the aid of special algorithms (25) to produce two-dimensional plots of skin depth and dielectric constant in the region between two drill holes. The difference between preshot and postshot plots shows promise of providing a useful map of explosive-induced fracturing.

During the course of the high-frequency measurements, it was observed that the direct signal and a reflected signal from the interface of the coal bottom and underburden produced interference effects (notching) in the received signal. The appearance of notching, as a function of frequency, is related to the distance from the antennas to the interface.

## EM Measurement of Fluid Permeability

Fluid permeability can be measured in situ by injecting an electrically contrasting fluid into the medium of interest and then measuring its diffusion as a function of the volume of injected fluid. To test this, a 3-m packer was placed in one of the abandoned shot holes and a transmitting antenna in a satellite hole, 4.6 m away from the shot hole. A receiving antenna was located in another satellite hole 4.6 m away from the shot hole and on the opposite side so that the transmitted signal had to pass through the volume enclosed by the packers (see Figure 8.16). A saline solution having a conductivity of about 10 times that of normal ground water was pumped at 4 liters/min into the shot hole. As the saline solution displaced the water in the coal, the transmitted signal was attenuated.

Because it is very sensitive and does not require the saline fluid to pass completely from the injection hole to another hole before data are obtained, this method will probably be very useful for measuring fluid flow distribution, which can in turn be used to obtain permeability distribution. By placing the transmitting antenna in the injection hole and positioning a number of receiving antennas at various depths in satellite holes, a three-dimensional presentation could be obtained.

These EM techniques can provide useful and important methods for characterizing anomalies in coal media and for determining fluid permeabilities.

**FIGURE 8.16: CONFIGURATION USED TO DETERMINE THE
FEASIBILITY OF EM MAPPING OF IN SITU FLUID
PERMEABILITY**

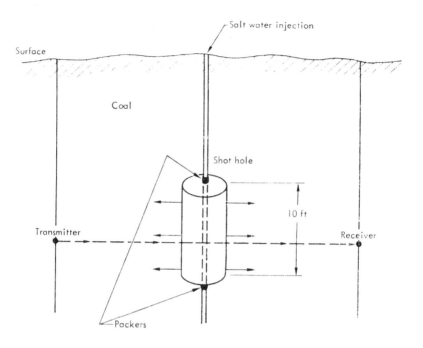

Source: UCRL-50026-75-1

**Coal Pyrolysis Measurements**

At LLL preliminary experiments on pyrolysis reactions of coal have been
undertaken. However, little quantitative information is available concerning the actual kinetics of subbituminous and lignite coal pyrolysis
and the effects of different atmospheric environments on the kinetics.

It is well known (28)(29) that by measuring product composition as a function of temperature at constant heating rate one can obtain information on
the rate of pyrolysis, and the associated activation energy and preexponential constant. Although this is not kinetic data in a rigorous sense, it is
sufficient to allow accurate modeling.

LLL designed and built an apparatus to study coal pyrolysis under different gaseous environments and heating rates. Figure 8.17 is a schematic
diagram of the apparatus. By using a heated sampling system with a flow

by-pass, it is possible to sample all products (gas and tar) without interrupting flow in the reactor. Since the flow of a carrier gas through the system is maintained constant ( ± 1%) throughout the experiment, a quantitative measure of gaseous product composition relative to the constant background of carrier gas is obtained. Furthermore, the carrier gas can be changed to simulate the pyrolysis environment of an in situ gasifier.

**FIGURE 8.17: EXAMPLE OF LLL COAL-PYROLYSIS APPARATUS**

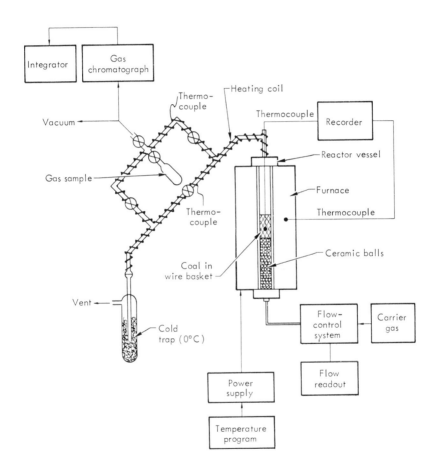

Source: UCRL-50026-76-3

**In Situ Continuous Temperature and Pressure Measurements**

Defining Temperature and Pressure Instrumentation Needs and Problems: The steps that must be taken to provide adequate temperature- and pressure-measuring equipment for the in situ processes are:

1. Determing the expected temperature and pressure ranges as well as other environments to which the measuring equipment may be exposed; determining which materials or protection may be necessary to combat the ill effects of environment on the measuring equipment.
2. Devising methods for emplacing the measuring equipment.
3. Selecting or designing temperature and pressure sensors that will reliably and accurately provide understandable data for process control; selecting or designing data reduction and control equipment that will translate the temperature and pressure information into control functions.

Temperature and Pressure Ranges and In Situ Environments: The burn front in the coal gasification process may attain 1500°C. The temperatures will approach the natural ambient of less than 100°C at increasing distances from the retorting front. Pressures will exceed 7 MPa in the coal project.

The coal gasification instrumentation will be exposed to corrosive environments that include hot water, tars, phenol, sulfur dioxide, hydrogen sulfide, methane, and carbon dioxide. In both processes, the earth overburden may move because of pressure buildup or subsidence. Large dynamic stresses and movements will also occur during the explosions used to fracture the coal or oil shale beds. All these factors must be taken into account in developing an instrumentation package.

The instruments that are to be used for in situ measurements must be made from materials that will withstand the process environments. If a protective shield is used to isolate the instruments from the reacting environment, the shield material must be constructed from resistant materials. The selection of the wrong materials can result in instrumentation failure because of corrosion or loss of instrumentation stability because of diffusion of environmental impurities into the temperature and pressure sensors. Therefore, extensive materials studies are necessary to the preparation of an instrumentation plan.

Instrumentation Emplacement: If instrumentation sensors and cabling are emplaced before explosive fracturing, they must be mechanically protected from the ensuing shock stress and earth movement. If the instrumentation is emplaced after fracturing, it may have to be in a cased hole. It may be expedient to route the instrumentation power and signal cabling from the bottom and along the side of the fractured zone.

The ideal emplacement method for temperature and instrumentation is a cased hole through which the temperature sensor can be lowered and raised from the surface. This not only gives accurate temperature distribution data but also facilitates maintenance and replacement of the sensor and its cabling.

Pressure transmission tubes can lead to the surface to give pressure distribution data. Unfortunately, using separately drilled and cased instrumentation holes, while desirable from the technical standpoint, is the most costly approach.

Sensing Devices, Data Reduction Equipment, and Process Controllers: Temperature Sensors — There are four standard types of thermocouples available to measure the temperature ranges of interest in the in situ processes: ISA types K, R, S and B. Choosing the thermocouples for in situ measurements requires that factors of maximum temperature, sufficient signal, and cost be balanced.

These thermocouples work relatively well in oxygen environments but are gradually changed by a wide range of impurities so that their signal does not reflect the actual temperature. Even thermocouples sheathed in metal and ceramics have shown degradation effects at long times at high temperatures.

Resistance thermometers of metallic construction are very accurate and produce large signal changes with temperature change. They are, however, usable only to 600° to 800°C. They also are degraded by mechanical deformation and shock loading. Certain high-temperature ceramics, however, change resistance with temperatures and probably could be developed into highly stable in situ sensors.

Magic Wire is a thermocouple-wire-based transducer that is embedded in a ceramic material and sheathed with a metallic covering. The ceramic has a negative coefficient of electrical resistance (i.e., it becomes more conductive with increasing temperatures). At high temperatures the thermocouple-wire pair shorts across the conductive ceramic, forming a temperature-measuring junction. A third wire is used to locate electrically the position of this short. This device not only could record burn front temperature but also could locate its position. The temperature limit of this material is 1100°C; it possesses most of the qualities of ISA type K thermocouples.

Electrical time domain reflectometry (TDR) can be used in a technique similar to the Magic Wire method. The setup is much the same as the

Magic Wire, except that the wires in the cable are not the thermoelements. As a result, the ceramic itself becomes the thermometer. Its resistance decreases with increasing temperature. TDR can be used to locate the point of lowest resistance (i.e., the point of maximum temperature). With this technique, the transit time of an electrical pulse is related to the distance to that point, and its amplitude is related to the resistance.

TDR can also be used to locate a set temperature. A TDR cable could be constructed of a metal that melts at a desired temperature. Upon reaching the melting point the wires would melt and fuse together. TDR then would be employed to find the location of this temperature point.

Acoustic thermometry exploits the change in sonic velocity that occurs with temperature. A metallic bar with a send-receive acoustic transducer mounted at one end can monitor the average transit time of a sonic pulse sent down the bar and reflected back from the far end. The average temperature of the bar can be calculated from the known velocity-temperature relationship for that metal. Temperature differences in the bar can be measured by placing notches at regular intervals along the bar. Each notch will reflect a part of the sonic pulse, and the times between reflected signals can be associated with the average temperatures between notches.

The accuracy of this type of device is limited by the accuracy of measuring the short time spans between notches. In addition, intergranular corrosion caused by harsh environments will severely distort the signal. A portion of the signal will be lost to the surrounding solids at points of contact, thus reducing signal magnitude.

Pressure Transducers: Pressure transducers are available commercially that can withstand the temperatures of the in situ processes. However, they are made of materials that would corrode severely in the chemical environment. They would also be damaged by mechanical shock and the forces of earth movements. These problems can be eliminated by bringing the pressure environment up to the surface. Making measurements at the surface would also eliminate the need for the more expensive temperature-resistant transducers.

Data Reduction Equipment and Process Controllers: This type of equipment has already been developed for surface industrial plants. Choosing the appropriate equipment for the in situ processes requires knowledge of expected signal levels, process actuators, process servomechanisms, and requirements for process continuation.

## Remote Optical Measurement of Temperature in the 0° to 2000°C Range

The effect of a birefringent crystal on the state of polarization of a transmitted light beam has been disclosed by LLL. The method has some unique attributes: remote nature of measurement, temperature resolution of better than 1°C, and high-temperature operation—up to the melting temperature of the sensor optics (2040°C for sapphire). A schematic of an idealized temperature measurement system is shown in Figure 8.18.

A linearly polarized light wave produced by a laser is directed to a birefringent crystal located in the thermal environment to be monitored. The orientations are such that there is a 45° angle between the optical polarization vector and the crystal optic axis. In this manner, two orthogonally polarized optical wavefields are created within the crystal, one parallel to the optic axis and one perpendicular. Upon propagation through the crystal, these superpose to create the transmitted light beam, the polarization state of which is generally different from that of the incident one.

### FIGURE 8.18: METHOD OF REMOTE MEASUREMENT OF TEMPERATURE

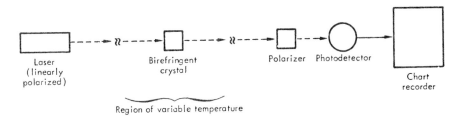

Source: UCID-16640

As the birefringence or the length of the crystal changes, i.e., due to temperature, the polarization state of the transmitted light beam also changes. The polarizer placed before the photodetector functions as an analyzer, converting these temperature-induced changes into a detectable intensity modulation that can then be displayed.

Applications: Field application requires a more sophisticated apparatus than that shown in Figure 8.18 in order to simplify alignment, increase reliability, and obtain data unambiguously. A step in that direction is shown in Figure 8.19. This system also offers increased sensitivity for a given crystal length because the TIR (total internal reflection) end doubles the effective optical path within the crystal.

**FIGURE 8.19: APPARATUS CONFIGURATION FOR FIELD APPLICATIONS**

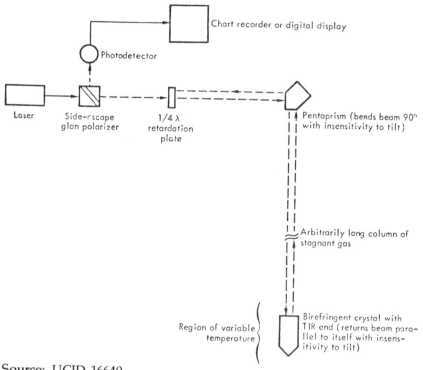

Chart recorder or digital display

Photodetector

Laser

Side-escape glan polarizer

1/4 λ retardation plate

Pentaprism (bends beam 90° with insensitivity to tilt)

Arbitrarily long column of stagnant gas

Region of variable temperature

Birefringent crystal with TIR end (returns beam parallel to itself with insensitivity to tilt)

Source: UCID-16640

Increased degrees of sophistication are possible, such as the use of split-field techniques developed for fringe-counting interferometry to indicate whether the temperature is increasing or decreasing.

The advantages of the method are:

1. Remote nature of the measurement. Using diffraction-limited lasers and optics, the temperature at the crystal can be measured at very great distances, either in turbulent environments using graded-index fiber optics (which preserve polarization properties), or through vacuum lines of sight.
2. High resolution. Preliminary demonstration experiments exhibited a temperature resolution of about 1°C (this assumes 1/10 cycle definition, which is not unreasonable). This sensitivity should be easily doubled by using a TIR crystal, and doubled again by decreasing the wavelength (i.e., argon-ion laser instead of He-Ne). In conjunction with sophisticated electronics, millidegree resolution seems feasible.

3. High temperature operation. The upper temperature range is limited only by the melting temperature of the optics.
4. Optical simplicity. The scheme can be designed to be alignment insensitive. It lends itself to array-type measurements where one laser can illuminate a number of crystals with high signal-to-noise ratio since fast response is not necessary.

The primary disadvantages include:
1. The necessity of a relatively tranquil line of sight between analyzer optics and the birefringent crystal.
2. The system response-time limitation imposed by the thermal-diffusion time constant of the crystal.


## MERC INSTRUMENTATION PROGRAMS

MERC, the Morgantown Energy Research Center in West Virginia is actively engaged in measuring quantitatively the intensive and extensive properties of in situ coal gasification. Described here are instrumentation systems projected for measuring, monitoring, and controlling the process.

### Field Instrumentation

For field monitoring, both surface and subsurface measurements are required to characterize and attempt to control the process. Best and most economical monitoring results are achieved if the monitoring wells are located in such a manner that the minimum number is used and that the least interference is caused by their presence. However, the gasification zone must be completely spanned if effective and thorough process control is to be exercised.

Effective control and evaluation of the process is dependent upon a good estimate of the burn front location and a map of the gasified region. Therefore, several measurement schemes are to be carried out, including thermal gradient and time rates, surface and subsurface resistivity, passive and active acoustic, and electromagnetic.

### Resistivity Measurements

Injecting a recurrent bipolar electrical current into the earth by means of a pair of emplanted electrodes and then measuring the potentials between many two-electrode point sets in the surrounding region can provide information useful for determining resistivity and irregularities in the earth's media based upon Laplace's equation. Figure 8.20 shows a tech-

**FIGURE 8.20: TRI-POTENTIAL ELECTRODE CONFIGURATION**

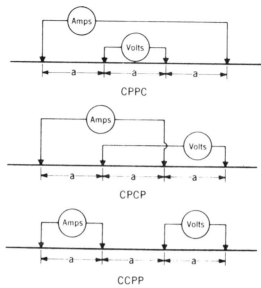

Source: MERC/SP-75/1

nique used at the MERC underground coal gasification site. The results of this tripotential method for verification of faults located by aerial survey and for providing background readings for the underground gasification of coal experiments can be illustrated. These basic methods can also be modified by using a large number of fixed potential probes at and below the earth surface as in Figure 8.21. The advantage is that simultaneous

**FIGURE 8.21: POTENTIAL MEASUREMENTS AT AND BELOW SURFACE**

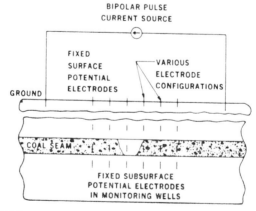

Source: MERC/SP-75/1

potential measurements from a fixed probe array allow automatic data collection and processing such as illustrated in Figure 8.22.

This pattern of current used as reference is significantly altered by the introduction of the gasification process anomalies such that detectable changes occur in the potential readings with time as the front progresses. Comparative potential measurements are taken psuedosimultaneously at time intervals from probe arrays on and below the surface. Probe-to-probe relative potentials for any combination can be selected in sequence by computer control which can be instructed to make adaptive decisions. A switch matrix connects the selected probe pair to a differential amplifier, the output from which is recorded on an analog tape recorder along with a coded timing signal that synchronizes the data gathering operation. Either a synchronous detector capable of some signal averaging or a sample and hold for instantaneous sampling can be used for input to an A/D converter. The resulting digital data can be examined, compared against previous values, inserted into mathematical simulators and, after processing, plotted on a visual display.

## FIGURE 8.22: RESISTIVITY MONITORING AND DATA REDUCING SYSTEM

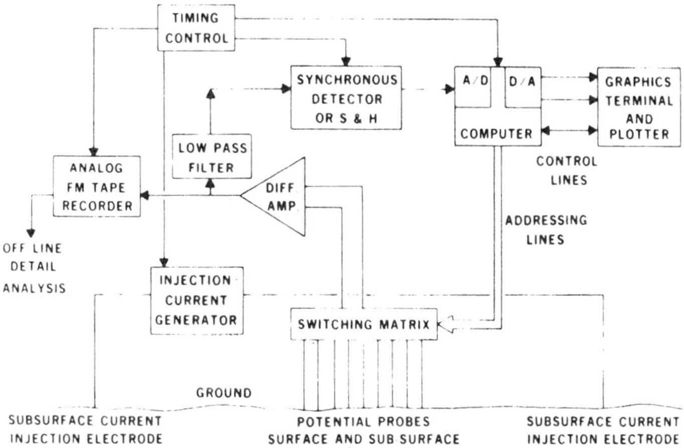

Source: MERC/SP-75/1

## Acoustic Systems

A second independent mapping approach is depicted in Figure 8.23 which presents an acoustic system for both passive and active modes of operation. Noises emanating from the activity associated with the burn progression serve to locate the front by triangulation if these disturbances are identifiable. Unwanted background from a multitude of sources necessitates sophisticated methods for recognizing useful signals. Some examples of these methods are threshold level detection, multi-channel coincidence, cross-correlation, power spectra and other time and frequency domain signature characterizations. Real time implementation of the more complex identification techniques require bulk storage such as multi-channel analog magnetic tape recording (continuous loop) or a much faster access memory media such as Charge Coupled Devices (CCD) memory for either analog or digital retention.

As the data become current, a graphics map is formed and updated, and at any point in time, a hard copy can be generated. Active mode diagnostics initiated by the activation of an explosive charge array (acoustic source) share the acoustic sensors of the passive mode. A reference pulse code is recorded on the magnetic tape recorder to mark the firing instant which is valuable in the subsequent off-line analysis procedure.

**FIGURE 8.23: PASSIVE AND ACTIVE ACOUSTIC MEASUREMENTS**

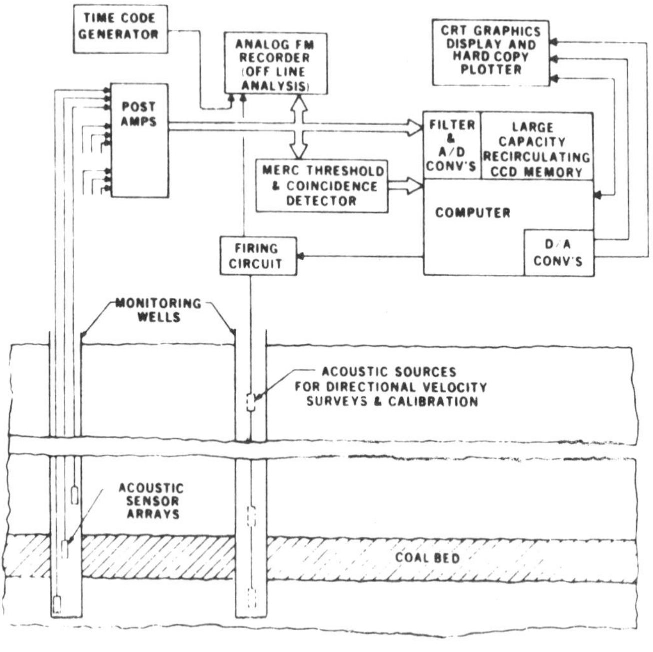

Source: MERC/SP-75/1

## Temperature Measurements

The heat generated in the gasification process will create time-varying thermal gradients that progress and fluctuate with the turbulence and intensity of the reaction zone activity. The time change in the temperature gradients offers a means of tracking heat flow and the levels indicate the proximity of the exothermic reactions. The combined temperature measurements can serve as a basis for mapping the reaction interfaces. Local 3-dimensional gradients and average gradients between all wells at several elevations in the well array will be acquired from thermocouple readings. A 3-D plot of the temperature and the predicted burning zone, arrived at by interacting the temperature data with a math model, can be exhibited on a CRT display with continual updating. The temperature field can also provide some information pertinent to the understanding of materials flow (tar) and water migration patterns. The thermal conduction waves have very low velocities, and there are sharp decreasing gradients ahead of the reaction zone, so stability of the temperature measuring system is important in detecting small temperature changes.

## Electromagnetic Radiation

Highly directional antenna arrays may be useful for electromagnetic diagnostic purposes at ranges not exceeding 50 feet into the coal bed at high frequencies. The frequencies used depend upon the attenuation that the electromagnetic radiation endures. The highest microwave frequencies generally show the greatest power loss in the coal bed. However, the size of an antenna array must be increased as frequency is decreased to maintain a fixed radiator RF beam width. This indicates that a compromise must be made within the constraints of size and shapes allowable within a borehole and well pattern. Figure 8.24 illustrates a directional electromagnetic radiation concept employing a subsurface antenna array that scans the coal seam to locate variations in the radiation impedance, and in the reflected and refracted components reaching the receiving array. Correlation of each is to be made with the location of the reaction, fluid flow interfaces and the temperature gradient peaks. The technique can be used to assess, prior to burn, the initial directional characteristics of the coal bed such as water channels, fracture and other directional features.

## Subsurface Gas Sampling

Evaluation of the chemistry of the process reactions, such as what reaction and the rates at which they occur at various field locations, necessi-

## FIGURE 8.24: ELECTROMAGNETIC SCANNING TECHNIQUES

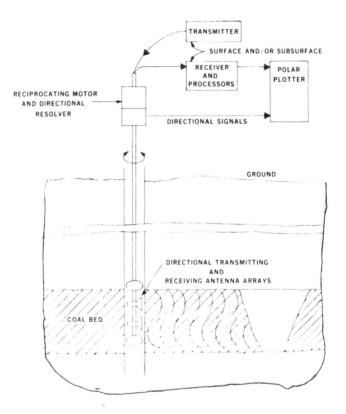

Source: MERC/SP-75/1

tates acquiring a thorough analytical knowledge of the gas composition throughout the region of interest. It is understood that although it is not physically realizable to accomplish this, certain useful information can be derived to formulate an approximate model by extracting gas samples from monitoring wells that penetrate the active region of the coal field and observing the process pass by the monitoring wells. Figure 8.25 illustrates one such sampling probe, which can be part of an encompassing array of probes. The sample collected through a permeable membrane (such as sintered stainless steel) is maintained as close as possible to its source temperature to reduce condensation by insulated tubing. Solids, water and tars are trapped ahead of the multiple collector mass spectrometer. A vacuum pump is included, if necessary, and a back flush compressor can be used to clear the sampling system. For complete gas condition assessment, both temperature and pressure are sensed at each sampling probe.

## FIGURE 8.25: SUBSURFACE GAS COMPOSITION MEASUREMENTS

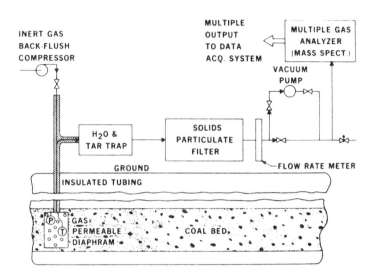

Source: MERC/SP-75/1

### Stress or Displacement Measurements

The gasification process is a strong function of the state of stress. Pressure and liquid and gas flow distributions are directly affected. Thus, in developing a commercial, progressive process, one must incorporate over-burden-induced stress as a primary consideration. The roof elevation must be known continuously and controlled, as in longwall mining, for efficient recovery.

Local relative and absolute lateral and vertical displacement and/or stress measurements in the vicinity of the coal bed can reveal a great deal about the process state and lateral extent. Along with structural creep and relaxation, thermal strain, permeability data, etc. evaluated by laboratory tests during various phases of gasification at known elevated temperatures, one can obtain much insight into the process when used in conjunction with structural finite element models.

Subsurface stress or displacement measurements are not easily made either. Axial displacement measurements in monitoring holes are perhaps the easiest and most reliable. Slide wire resistive elements whose resistance is a known function of displacement is one means of measurement.

## Surface and Subsurface Water Monitoring

The monitoring of surface and subsurface water is necessary, largely for environmental reasons. However, water quality measurements can also be useful in analyzing in situ gasification conditions. In some cases, collecting water samples for laboratory analysis is adequate, but the technique is not compatible with the general objectives of real time process monitoring. Even for environmental purposes, a quick response could prevent excessive damages. Numerous instruments exist for continuous water quality monitoring. The problem is monitoring a particular characteristic with the background or other conditions present from the in situ gasification process.

Perhaps the first order detection type measurement to be made on surface and subsurface water is electrical resistivity. This measurement is relatively simple in comparison with many others and represents a broad spectrum of changes which could be introduced by the gasification process. More detailed types of measurements being considered beyond the initial resistivity changes include total dissolved solids, dissolved oxygen, acidity, alkalinity, phenols and other select mineral or compound tests.

## Air Quality Monitoring

The atmosphere must be monitored continuously at the site for any toxic contaminants released by the surface or subsurface gasification system. An eight component mass spectrometer with four input stream selections can be shared for this service. Separate carbon monoxide and nitrogen oxide detectors provide PPM measuring capability for sampling the atmosphere inside trailers which house instruments and operating personnel.

## Instrument Subsurface Integration and Packaging

The instrumentation to be emplanted in the monitoring well array may require cementing or grouting in place, if adequate coupling to the local medium cannot otherwise be obtained, or if the effects of surface oriented disturbances cannot be eliminated in some other manner. Many of the measuring systems described will share the same set of wells. The problem of compatability must therefore be considered in the integration of these devices into total packages. Electrical, mechanical, acoustic, EM and thermal cross interference all contribute to a complex integration problem that is aggravated by the extreme conditions and restraints imposed by the process and the well geometry.

## Surface Process Instrumentation

Commercialization of in situ gasification would be greatly enhanced by exclusive surface measurements. All input and output energy and materials flow must be inventoried, as completely as is reasonably possible, to determine mass and energy balances, etc., for evaluating process performance with respect to gas quality, production rates, efficiency, resource recovery, feedforward and feedback process control, and for environmental considerations. Instruments such as pressure and flow transmitters, temperature sensors, gas, liquid and solid analyzers, liquid and solid weighing devices must be used extensively.

## Data Acquisition and Control Systems

Laboratory experiments are used to develop measurement techniques, evaluate process control schemes, develop computer simulation models for process control and optimization, determine the monitoring facilities required to support real-time computer process control models for field operations, and supply data for mathematical models. Data acquisition and process control hardware and software developed for these lab experiments can be modified and expanded for field applications. The general capabilities are illustrated in Figure 8.26. The variables monitored include: gas and liquid flow rates, gas compositions, temperatures, pressures, stresses and strains, resistivity, acoustic emission and acoustic velocities.

One of the reactor simulation vessels installed has over 40 access ports used to measure the different variables and take gas samples radially and longitudinally along the axis of the hole through the coal sample. The instrumentation response, data reduction and display time for most variables under these laboratory conditions is approximately one second. The laboratory experiments are conducted under well defined conditions for studying the reaction kinetics phenomena. Experimental data on gas composition, calorific values and temperatures are compared with data generated from theoretical model solutions obtained for the same boundary conditions. The boundary conditions are those of the stream gasification method which consists of, in this case, a through hole in a block of coal and uniform ignition initially along the hole.

## FIGURE 8.26: MERC DATA ACQUISITION AND CONTROL SYSTEMS

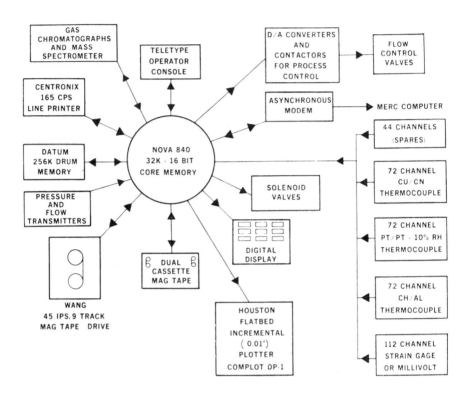

Source: MERC/SP-75/1

## SANDIA LABORATORIES INSTRUMENTATION STUDIES

The environments associated with in situ processes are characteristically high temperature, high pressure, and corrosive. These characteristics, combined with the inaccessibility of location, pose a formidable task for the development of reliable, accurate, cost-effective, and long-life instrumentation systems. The definition and control of the flame front during in situ coal gasification may well be the most complex application of an instrumentation system. Thus, in situ coal gasification serves as an appropriate example for the various instrumentation techniques, applications, and problems.

**Flame-Front Position and Movement**

The continuous knowledge of the position and movement of the flame or combustion front in a coal gasification experiment is considered to be critical to the optimization of in situ coal gasification. Possible instrumentation techniques which can detect the discontinuity between unreacted coal and the void produced behind the front include those based upon acoustic wave propagation (30)(31)(32)(33)(34), very low frequency (VLF) electromagnetic wave propagation (35)(36)(37), and detection of gravitational anomalies.

Acoustic holography is potentially capable of giving a three-dimensional representation of a subsurface anomaly. The technique is implemented by placing an array of geophones on the surface or in neighboring boreholes in the vicinity of the volume to be mapped and then exciting the region with sound waves produced by an explosion or a mechanical thumper. Resolution depends upon the use of a large number of geophones. The array diameter (aperture) must be several times the depth of the feature being investigated, and the fineness with which the aperture must be gridded with transducers is determined by the wavelength of the sonic frequency being detected. In no case can the geophones be spaced further apart than one-half this wavelength, and a spacing closer by a factor of two or three is desirable. Useful wavelengths are determined by the transmission properties of the underlying soil which, in general, tend to attenuate shorter wavelengths. For depths on the order of a mile or more (typical in petroleum exploration), the shortest usable wavelength is on the order of 500 feet. Closer to the surface, wavelengths on the order of 50 feet might be used. In that resolution varies with the wavelength, it is desirable to use the shortest practical wavelength with the attendant greater density of geophones. However, the total number of geophones can be reduced because the measurement can be made piecemeal by gridding part of the aperture at a time. Moreover, with this technique, unwanted background details are suppressed by a method of subtraction and only changes which have occurred are scrutinized; thus, the overall data acquisition and analysis effort is reduced.

The holographic output may be in the form of a hologram transparency which can be optically processed to reconstruct an image, or it can take the form of a computer tape which is to be further processed and then drawn by machine graphics. In either case, the final product is a three-dimensional representation of the subsurface feature and its changes in shape and position as a function of time.

Very-low-frequency (VLF) electromagnetic wave propagation also may be useful in this application. The scheme is analogous to conventional eddy current techniques but at considerably lower frequencies (1 to 50 Hz).

Two general techniques have promise. The first is to locate both the transmitting and the receiving coils on the surface and measure the coupling through the earth. Changes in resistivity brought about by reaction and mass removal would be measured. A second technique is to locate the transmitting and receiving coils in situ in an attempt to precisely locate and obtain details at the reaction zone. It seems reasonable that the hot gases at the reaction zone have a resistivity different from the coal which has a resistivity different from the surrounding overburden. These differences should be detectable either from the rear of the reaction or from its front. The rear of the reaction zone appears to be favorable because the incoming air could be used to cool the coils and because the reaction residue may fall enough to provide a solid-free path to the reaction. The coils would be large, several feet in diameter, and therefore would probably have to be extended after they are positioned underground.

A gravimeter, a device for measuring the local acceleration of gravity can be used to detect changes in the mass of material within several hundred yards of the instrument. This, in general, would be an average measurement of the total mass extracted from an in situ process; however, the measurement could be made continuously to monitor the extraction process. The sensitivity of present cable-supported, borehole gravimeters is 3 μgal (1 gal equals $1 cm/sec^2$) which is marginal for these applications in that it corresponds to the extraction of approximately 1,500 cubic yards of coal. Laboratory gravimeters for tidal measurements are capable of thirty times greater sensitivity, which would imply detection of 50-cubic yard extractions. However, improvements in borehole gravimeter sensitivity can be made; and in addition, the use of multiple gravimeters which are solidly emplaced in adjacent boreholes would improve overall sensitivity and enable temporal and spatial measurements of the mass change.

## Chemical Analysis

Measurements of the presence and concentrations of chemical species at or near a reaction zone are important in the investigation and understanding of an in situ experiment or process. Ultraviolet and visible spectroscopy based upon electronic transitions and infrared spectroscopy based on vibration-rotation transitions are used to detect and measure the concentrations of molecular species. While these methods and their instrumentation are well developed in the laboratory, they have to be adapted for these in situ environments. One viable approach would use refractory light pipes to probe near the reaction site and transport the light to a remote site for dispersion and detection. Suitable high temperature materials for ultraviolet, visible, and near infrared light pipes are quartz and aluminum oxide. For obtaining absorption spectra, the possibility of using a tunable dye laser to provide a monochromatic source of high specific intensity is promising. Again, it is proposed that the laser be located as close as tolerable to the point where measurements are to be made and that the remaining distance be covered by means of the refractory light pipe technique.

Another technique which appears promising is the use of specific ion electrodes based on refractory solid electrolytes to determine concentrations of reactive species such as oxygen, hydrogen and sulfur (38). Devices such as those envisaged have been widely used in the laboratory at lower temperatures. Their application at high temperatures also has precedent in steel making where they are used to determine concentrations of sulfur in metals and slags (39). The basic theory for these devices and some materials suitable for use in high temperature, remote, and adverse environments are available for development of this technique. The chief problems will be to extend the upper temperature limits of existing probes and to develop new probes capable of analyzing species other than oxygen, hydrogen, and sulfur.

Gas analyses by mass spectroscopy and gas chromatography are also required to monitor product gas streams. This can be done above ground, and the techniques are well developed. However, the development of a rugged, miniaturized quadrupole-type spectrometer would have definite application by allowing gas compositions to be determined as near the reaction front as temperature permits. A number of these emplaced in separate boreholes could provide additional spatial information. Sandia Laboratories has developed such sensitive analytical instrumentation for rocket and balloon atmospheric sampling studies (40)(41)(42).

## Instrumentation Package

Instrumentation packages which would be located on a grid of accurately located boreholes could be combined with sophisticated data processing and analysis equipment to provide an effective method of obtaining information from an in situ process. Comprehensive instrumentation packages can be developed, although the exact components of the packages would depend upon the specific application, degree of desired sophistication, and number of instrumentation channels available. Each of the individual transducers within a package would provide continuous data. When these data are combined with data from similar transducers in other packages and with data associated with other parameters, a comprehensive description of the in situ process is potentially available. Further, such continuous monitoring will definitely be required for the subsequent control and   optimization of the process.

The following package appears attractive for an in situ gasification process: thermocouples, total pressure transducer, quadrupole gas analyzer, geophones, flowmeters, and necessary data processing equipment. Such packages at known locations would enable additional diagnostic techniques, such as tracer methods, to be implemented. A selected isotope or inert gas could be released at one hole, and times-of-arrival and concentration profiles obtained at the other locations would give information on relative permeabilities and extent of reaction. Abrupt changes in permeability detected at a given location might indicate the onset of channeling or bypassing in the gasification process. In addition, quantitative gas analyses will provide a convenient and continuous measure, via mass balance calculations, of the amount of coal that has been extracted.

## Data Transmission

Much of the diagnostic data is to be acquired from transducers emplaced deep underground and, in addition, communication with emplaced devices may be required for control applications. Maintenance of signal integrity in the presence of high temperatures, pressures, and corrosive gases will be a difficult task. Thus, both hard-wire (cable) and remote (electromagnetic and acoustic) data transmission systems will be considered. The less equipment, such as radio frequency (RF) transmitters and analog-digital (A/D) converters, emplaced underground, the lower will be the failure rate and cost. There will be compromises, because the transmission of low-voltage signals over long distances in noisy environments results in low signal-to-noise ratios.

A survey of applicable work within Sandia Laboratories indicates that cables can be designed to survive maximum temperatures of 1000°C; however, continued exposure to lower temperatures will also pose design problems. Previous development of a design utilizing boron nitride fibers interwoven around a ceramic material has demonstrated cable survival to 800°C (43)(44). This development was conducted on short cable lengths; more effort would be required to develop the longer, higher temperature cables required for this program. Other insulation systems such as carbon and polyimide combinations (45) and intumescent organic foams, which are presently used for protection in fire environments, should also be investigated. Further, Kapton insulated cables capable of surviving 300°C have been developed (46). Modifications of these designs will be studied to determine whether their survival time is adequate for some applications.

Simpler cable designs, encompassing bare conductors passing through perforated ceramic spaced at regular intervals, appear capable of design to these high temperatures. Cables can also be routed away from the combustion zone so that lower temperature cable designs will suffice.

Both electromagnetic and acoustic wave propagation through the earth appear feasible in cases where direct cable transmission is not possible. The attenuation of electromagnetic waves by the earth is severe and is affected by the medium's conductivity and the wave frequency. The transmitter power required for a given communication link is proportional to the cube of bandwidth and sixth power of range and is inversely proportional to the effective volume of the antennas. However, it has been found that this technique is useful for limited range data telemetry in several programs (35)(36)(37)(47). A 10-watt transmitter and a 6-inch-diameter by 6-foot-long solenoid antenna were used to transmit three channels of 100-Hz data through 355 feet of dolomite rock. A 90-watt system with solenoid antennas was designed to transmit a single 100-Hz channel about 1,000 feet through earth of higher conductivity. During Plowshare Tests, 5-Hz and 10-Hz signals were transmitted from a 65-watt transmitter and 1,200-foot-radius loop antenna on the surface to a receiver 11,000 feet deep inside a cased borehole.

These experiences indicate that it is feasible and practical to transmit low-frequency data through solid media. Such systems should be designed to optimize system parameters for a given transmission path. Medium properties could be measured with standard oil-well logging techniques. In addition, preliminary transmission tests will be required to define and verify the transmission model. An appropriate thermoelectric power source to energize such a system for long-term measurements will also be required.

The feasibility of transmitting elastic waves (an acoustic or seismic data link) through the earth to depths of 15,000 feet has been investigated (33)(34). This work has concentrated on transmitting low-frequency signals ($<50$ Hz) from the surface and on studying the attenuation effects versus depth and frequency. A single vibrator truck has been used to obtain adequate signal-to-noise ratios to depths of 8,500 feet. Greater depths appear to require more than one source because of a limit to the amount of energy that can be coupled elastically into the earth. In a communication system, constant-frequency seismic signals would be utilized in a frequency shift keying scheme to transmit the desired information. Relatively portable electromechanical transducers have been developed which could be emplaced in a confined volume. Characterization of the range of these transducers would be required; and, more than likely, a relay network would be required for a deep (>3,000 feet) process. If the transducer were coupled directly to a well casing, the losses could be considerably less (particularly if the casing were not attached to the media) and the relay network might not be required.

### Data Acquisition, Computer Analysis, Feedback and Control

Data acquisition, handling, and computer processing; data analysis; and data display are significant elements of a total instrumentation system for the measurement and control of an in situ process. However, the exact configuration of a given system and its components will largely be determined by the application requirements. Some modification of Sandia Laboratories' remote experiment monitoring system for underground nuclear testing is probably applicable to in situ coal gasification or other processes (45)(49)(50)(51)(52)(53)(54).

A schematic diagram of the system is shown in Figure 8.27. The system has been designed to monitor the status of various analog and discrete digital (on-off) functions measured in an underground experiment. The numerous remote area encoders, each capable of monitoring six analog and eight discrete inputs, are inexpensive, rugged packages with signal conditioning. Up to fourteen of the remote encoders operate asynchronously into a single multiplexer which can be located up to 2,500 feet away. Each multiplexer provides a single asynchronous pulse code modulated (PCM) bit stream for transmission to a data center either by up to 5,000 feet of coaxial cable or microwave. A number of these multiplexers can operate at the same time at various locations at the experiment site. At the data center, asynchronous demultiplexers (same number as multiplexers employed) reconstruct the PCM bit streams. These individual bit streams are presented to an interface-data preprocessor on the input channel to a medium-sized computer. Present system capability is 336 analog inputs (true differential), 448 digital discrete inputs at data rates from 2,400 bits/sec to ~10 megabits/sec (55).

A digital computer (such as a medium-speed minicomputer) is used for data processing and as a controller. Data from the experiment are processed in the computer and then displayed, stored, or transmitted to another location. It is imperative to have concise data displays so that either human or automated monitors can maintain a constant picture of

**FIGURE 8.27: EXAMPLE OF AN UNDERGROUND EXPERIMENT
MONITORING SYSTEM**

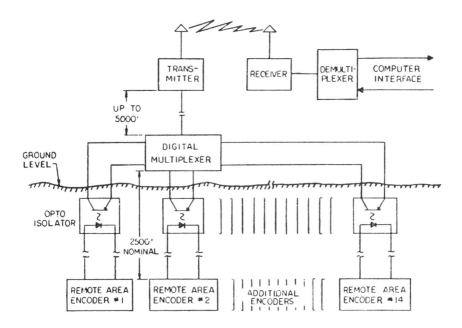

Source: SLA-73-0919

the process. An operator can change the inputs to the process, or the computer can be programmed to automatically control the process if a preprogrammed set of control and tolerance conditions are stored in the computer. The modular configuration shown in Figure 8.28 shows the overall data system and the various output and interaction functions that are available.

The proposed system is flexible and can be configured to whatever inputs, measurements, and outputs are required. In that individual modules and instrumentation trailers for this application are available, much long term development effort and expense should be eliminated. Also, a number of the trailers contain low-frequency multiplex signal conditioning and data recording systems as well as the usual complement of higher frequency magnetic tape and scope recording systems. These recording facilities and trained operating personnel would be immediately available.

# FIGURE 8.28: MODULAR CONFIGURATION OF AN OVERALL COMPUTER-CONTROLLED DATA ACQUISITION, ANALYSIS, AND CONTROL SYSTEM

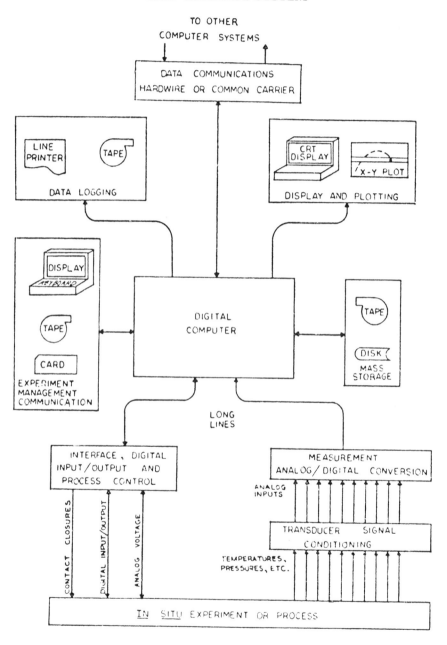

Source: SLA-73-0919

### Environmental Monitoring

An important argument for the in situ conversion of energy resources has been that many potential environmental problems such as sulfur emissions, strip mining, and water pollution would be alleviated. However, some processes will pose a problem if the process is not kept within operational bounds or under surveillance; the accidental venting of an in situ coal gasification process would be an example of such a hazard. Infrared scanning and remote air sampling are two examples of instrumentation systems which could be applied to an environmental monitoring aspect of an overall instrumentation systems development program.

The surface in the vicinity of an underground process can be monitored by infrared (IR) scanning in order to detect temperature changes symptomatic of venting. The fundamental tool would be a scanning IR camera sensitive to radiation in the 8 to 14μ region; this radiation window, which is above solar radiation, has proved successful in earth scanning. In such scanning applications, these cameras can detect temperature differences smaller than 0.5°C and resolve areas less than 1 ft² under known conditions. A reference survey would be made prior to initiating the coal reaction, and subsequent surveys would be made at regular intervals to detect any local heating of the surface that might indicate potential venting. The factors which affect the scan frequency are [1] the type of venting postulated, based upon experience and geology; [2] the cost of an unchecked vent; and [3] the cost of surveillance.

Development is required in the determination of the scan area and in the interpretation of the resultant scanning images. The monitored area is determined by the sensitivity of the IR camera, its normal viewing angle, the extent of the in situ process, and the local geology. These factors would determine the number of cameras and type of platforms which would be required. More complex is the interpretation of the data. The apparent temperature of an object as determined by an IR camera is a function of its emissivity, and factors that change a site's emissivity change its apparent temperature. Thus, the site's emissivity will undergo daily and seasonal changes and will be affected by most meteorological conditions such as rain, snow, or clouds.

Monitoring of explosive, toxic, and radioactive gases has been performed previously by Sandia Laboratories. The problems of remote gas sampling from a few thousand feet to 25 miles have been satisfactorily solved, and the types of sensors used have changed with the development of new equipment as have the procedures of calibrating and using these sensors. These detector systems have operated remotely with data transmitted via telephone, dataphone, hardwire direct, hardwire multiplexed, and RF multiplexed. The detectors have been developed and constructed to withstand severe environments of shock, vibration, moisture, dust, and temperature.

A number of sensor types are used for measuring radiocative gases; a Neher-White ion chamber calibrated for six-decade dynamic response of 1 to $10^6$ mR/hr has been used extensively. A Johnson-Williams type of sensor was selected for the measurement of toxic and explosive gases. The Percent-Explosive-Gas (PEG) system included a detector and power supply placed in the gaseous environment to be measured and a readout indicator placed at a remote location. A gas chromatograph system utilizing long-line gas sampling techniques might be useful in the analysis of gases present at an in situ process (56)(57). Thus, high-quality laboratory-type instruments can be placed up to 7,000 ft from a hostile environment. Equipment and procedures in use have been modified so that qualitative and quantitative measurements of a wide variety of gases can be made. These gases include $H_2$, $O_2$, $CH_4$, CO, $CO_2$, $C_2H_6$, $SO_2$, NO, $NO_2$, and $H_2S$; ammonia and acid gases could also be determined if required. The level of detection is a function of a number of factors, but is typically 100 ppm for hydrogen and methane.

## REFERENCES

(1) Gordon W. Greene, R. M. Moxham, and A. H. Harvey, *Aerial Infrared Surveys and Borehole Temperature Measurements of Coal Mine Fires in Pennsylvania*, Proc. VI Symposium Pennsylvania Society of Engineers (1969).
(2) R. Loison, A. Peytary, A. F. Bower, and R. Grillot, "The Plastic Properties of Coal," *Chemistry of Coal Utilization*, Supplementary Vol., H. H. Lowry, ed. John Wiley & Sons, Inc., New York, 1963, pp. 150-201.
(3) A. S.T. M., "Standard Method of Test for Free-Swelling Index of Coal, Designation: D720-67," *1973 Annual Book of ASTM Standards, Part 19, Gaseous Fuels; Coal, and Coke,* pp. 85-91.
(4) J. A. Dulhunty and B. L. Harrison, "Some Relations of Rank and Rate of Heating to Carbonization Properties of Coal," *Fuel, 32* (4), 441-450 (October 1953).
(5) M. Mackowsky and E. Wolff, "Microscopic Investigations of Pore Formation During Coking," *Coal Science*, Advances in Chemistry Series No. 55, R. F. Gould, Ed. American Chemical Society, Washington, D. C. (1966).

(6)  M. Mackowsky and W. Wolff, "Coking Properties of Coals of Different Grades with Special Regard to Bulk Density and Rate of Heating of Charge," *Erdöl und Kohle-Erdgas-Petrochemie, 18* (8), 621-625 (August 1965).

(7)  Y. Toda, "Densities of Coals Measured with Various Liquids," *Fuel, 51,* 108-112 (April 1972).

(8)  Y. Toda, "A Study by Density Measurement of Changes in Pore Structures of Coals with Heat Treatment: Part 1. Macropore Structure," *Fuel, 52,* 36-40 (January 1973).

(9)  Y. Toda, "A Study by Density Measurement of Changes in Pore Structures of Coals with Heat Treatment: Part 2. Micropore Structure," *Fuel, 52,* 99-104 (April 1973).

(10) R. L. Wong, *Coal Swelling Test,* Internal memorandum, Lawrence Livermore Laboratory (January 29, 1975).

(11) R. Vetterlein, *Coal Shrinkage Tests,* Internal memorandum, Lawrence Livermore Laboratory (February 13, 1975).

(12) A.S.T.M., "Standard Method of Test for Plastic Properties of Coal by the Constant-Torque Gieseler Plastometer, Designation: D2639-71," *1973 Annual Book of ASTM Standards, Part 19, Gaseous Fuels: Coal and Coke,* pp. 385-390.

(13) N. Y. Kirov and J. N. Stephens, *Physical Aspects of Coal Carbonisation* (Kingsway Printers Pty., Limited, Caringbah, Australia, 1967).

(14) D. W. van Krevelen, F. J. Huntjens, and H. N. M. Dormans, "Chemical Structure and Properties of Coal NVI-Plastic Behavior on Heating," *Fuel, 35* (4), 462-475 (October 1956).

(15) J. N. Stephens, *The Calibration of the Gieseler Plastometer in Terms of Absolute Viscosity Units,* C.S.I.R.C. Fuel Research Report M130, Australia (June 1957).

(16) L. A. Schrider and J. W. Jennings, *An Underground Coal Gasification Experiment, Hanna, Wyoming,* SPE 4993, presented at the 49th Annual Meeting of the SPE of the AIME, Houston, Texas, October 6-9, 1974.

(17) R. Van Nostrand and K. Cook, *Interpretation of Resistivity Data,* U.S. Geological Survey Prof. Paper #499, 1966.

(18) R. P. Reed, *Branched Thermocouple Circuits for Research on Energy Resource in In Situ Conversion,* SAND 74-0395, Sandia Laboratories, Albuquerque, New Mexico, 1975.

(19) L. A. Schrider, *The Hanna Experiments,"* presented at the First Annual Underground Coal Gasification Symposium, Laramie Energy Research Center, Laramie, Wyoming, July 30, 1975.

(20) D. R. Stephens, *In Situ Coal Gasification,* Lawrence Livermore Laboratory, UCRL-75494, January 31, 1974.

(21) D. W. Gregg, *Critical Parameters of In Situ Coal Gasification,* Proceedings of the 15th Annual ASME Symposium on Resource Recovery, Albuquerque, New Mexico, March 6-7, 1975.

(22) R. J. Lytle, et al., *The Lisbourne Experiment: Propagation of HF Radio Waves Through Permafrost Rock,* Lawrence Livermore Laboratory, Livermore, Rept. UCRL-51474 (1973).

(23) R. J. Lytle, *The Yosemite Experiments; HF Propagation Measurements Through Rock,* Lawrence Livermore Laboratory, Livermore, Rept. UCRL-51381 (1973).

(24) R. J. Lytle, R. Jeffrey, E. F. Laine, and D. L. Lager, *Coal Fracture Measurements Using In Situ Electrical Methods: Preliminary Results,* Lawrence Livermore Laboratory, Livermore, Rept. UCID-16639 (1974).

(25) D.L. Lager and R.J. Lytle, *Computer Algorithms Useful for Determining a Subsurface Electrical Profile,* Lawrence Livermore Laboratory, Livermore, Rept. UCRL-51748 (1975).

(26) J. W. Sherman and J. W. Woods, *Acoustic Array Methods for Instrumentation of In Situ Coal Gasification,* Lawrence Livermore Laboratory, Livermore, Rept. UCID-16591 (1974).

(27) A. B. Macknick, *Pyrolysis of Subbituminous Coal,* Lawrence Livermore Laboratory, Rept. UCID-16710 (1975).

(28) H. A. G. Chermin and D. W. van Krevelen, *Fuel 36,* 85 (1957), and references cited therein.

(29) H. Juntgen and K. H. Van Heek, *Fuel 47,* 103 (1968), and references cited therein.

(30) G. L. Fitzpatrick, H. R. Nicholls, and D. R. Munson, *Experiment in Seismic Holography,* U.S. Department of Interior, Bureau of Mines.

(31) R. A. Baker, D. E. Bishop, and G. C. Stoker, "Ultrasonic Characterization and Computerized Analysis of Graphite Billets," *Inter. Jour. of Nondestructive Testing,* 4, 1973, pp. 301-341.

(32) R. E. Spalding, *Unattended Seismological Observatory,* Sandia Laboratories, Albuquerque, New Mexico, SC-M-69-403, June 1969.

(33) D. B. Starkey, *Seismic Control System: Feasibility Tests,* Sandia Laboratories, Livermore, California, SCL-DR-70-86, November 1970.

(34) D.B. Starkey, *Seismic Transmission Tests: Vibratory Source,* Sandia Laboratories, Livermore, California, SLL-73-0038, August 1973.

(35) K. B. Kimball, *Subterranean Radio Frequency System for Earth Motion Data Acquisition,* Sandia Laboratories, Albuquerque, New Mexico, SC-DR-70-463, 1970.

(36) D. B. Starkey, *Electromagnetic Transmission and Detection at Great Depths,* Sandia Laboratories, Livermore, California, SLL-73-5278, 1973.

(37) R. J. Tockey, *On the Feasibility of Electromagnetic Communication Through the Earth for Command and Control of Sequentially Fired Plowshare Explosives for Gas Well Stimulation,* Sandia Laboratories, Livermore, California, SLL-73-5250, August 1973.

(38) C. Wagner, *Proc. Int. Comm. Electrochem. Thermo. and Kinetics,* 7th Meeting, London, 1955, Butterworth Scientific Publications.

(39) W. M. Boorstein, R. A. Rapp, and G. R. St. Pierre, *High-Temperature Electrochemical Research in Metallurgy,* Air Force Materials Laboratory, Dayton, Ohio, AFML-TR-73-67, April 1973.

(40) T. C. Looney, *A Densitometer for Research in the Upper Atmosphere,* Sandia Laboratories, Albuquerque, New Mexico, SC-DR-70-616, November 1970.

(41) R. O. Woods, *Optical and Charged Particle Instrumentation for Auroral Experiments,* Sandia Laboratories, Albuquerque, New Mexico, SCTM-72-0536, August 1972.

(42) G. C. Tisone, R. O. Woods, and J. M. Hoffman, *Rocket-Borne Auroral Optical Measurements,* Sandia Laboratories, Albuquerque, New Mexico, SC-RR-72-0543, August 1972.

(43) Cecil L. Page, *Filament Wound Electrical Insulation Boron Nitride Fibers Interwoven Around an Ablative Material,* Sandia Laboratories, Albuquerque, New Mexico, SC-RR-67-3022, April 1968.

(44) Robert T. Sylvester, *Process Description for High Temperature, High Voltage Shielded Flat Electrical Cable,* Sandia Laboratories, Albuquerque, New Mexico, SC-TM-72-0780, December 1972.

(45) Edward A. Salazar, and Richard K. Traeger, *Aging of Polyimide Films and Cable Construction,* Sandia Laboratories, Albuquerque, New Mexico, SC-TM-71-0166, October 1971.

(46) R. T. Sylvester, *Characterization of Kapton Dielectric Strength and Insulation Resistance at Elevated Temperatures,* Sandia Laboratories, Albuquerque, New Mexico, SC-TM-71-0332, July 1971.

(47) T. W. H. Caffey, *Fresh Water Propagation Tests,* Sandia Laboratories, Albuquerque, New Mexico, SC-DR-65-459, 1965.

(48) L. J. O'Connell, *Demonstration of Isolated Analog Voltage Control,* Sandia Laboratories, Albuquerque, New Mexico, Laboratory Report 1812-73-1, March 1973.

(49) L. J. O'Connell (Coordinator), *Proceedings of Minicomputer Applications Symposium,* Sandia Laboratories, Albuquerque, New Mexico, SC-M-71-0055, February 1971.

(50) M. E. Daniel, and P. F. Martinez, *Minicomputers, Software, and Automated Testers,* Sandia Laboratories, Albuquerque, New Mexico, SC-DR-72-0080, February 1972.

(51) M. E. Daniel (Coordinator), *Proceedings of Measurements and Instrumentation Symposium,* Sandia Laboratories, Albuquerque, New Mexico, SC-M-70-851, December 1970.

(52) W. K. Paulus, R. T. West, and T. J. Gill, *Data Acquisition and Process Control Automation of System Environmental Test and Shock Simulation Facilities Feasibility Study,* Sandia Laboratories, Albuquerque, New Mexico, SC-DR-72-0387, January 1973.

(53) B. Stiefeld, *Computer-Based Display of Nondestructive Evaluation Data,* Sandia Laboratories, Albuquerque, New Mexico, SC-R-71-3405, October 1971.

(54) A.B. Campbell, "Computerized PCM Data Presentation and Real Time Monitoring Scheme," *Telemetry Journal 7,* p. 11, 1972.

(55) J. L. Rae, and R. Minter, *TC721, Product Specification, Remote Equipment Monitoring Systems,* Sandia Laboratories, Albuquerque, New Mexico, PS-T00899 Issue A, March 1973.

(56) L. W. Brewer, and D. R. Parker, *Gas-Chromatographic Analysis of Gases Found in Post-Shot Tunnel Systems,* Sandia Laboratories, Albuquerque, New Mexico, SC-RR-71-0337, June 1971.

(57) D. R. Parker, and L. W. Brewer, *Gas Sampling Through Lone Lines,* Sandia Laboratories, Albuquerque, New Mexico, SC-DC-70-5186, October 1970.

# International Developments

The material in this chapter was excerpted from UCRL-50026-75-4, UCRL-Trans-10867, and *Chemical and Engineering News*. For a complete bibliography, see p 251.

## WESTERN EUROPE

According to an article in *Chemical and Engineering News* of Dec. 6, 1976, Belgian and West German scientists are teaming up to study the feasibility of gasifying coal deep underground. The tests are expected to last at least three years. If successful, the technique could open up an important energy source for all of Western Europe.

But there are broader implications to the experiments being undertaken by Belgium's Institut National des Industries Extractives (INIEX) and West Germany's Kernforschungsanstalt (KFA). By substituting hydrogen for air in the underground combustion process, it could provide a means of generating substitute natural gas directly from the coal deposits.

Western Europe still is very rich in coal. It is thought that there is at least 200 billion metric tons of coal at a depth of about 2,000 m in the region stretching from the Ruhr area in West Germany through northern Belgium and into the Netherlands; there likely is still more at greater depths. The problem of putting it to use is an economic one.

Underground gasification is being looked at in a number of countries. In the U.K., for instance, the National Coal Board has been evaluating the technique, although the urgency of the study has been lessened somewhat by the discovery of sizable quantities of oil and natural gas in the British sector of the North Sea. The Soviet Union is farthest along in the practical application of the technology with four or five electricity generating stations powered by gas produced underground. A summary of their efforts is in the next section.

In Belgium's case, coal was the primary energy source until fairly recently. Some 15 or 20 years ago about 30 million metric tons were mined there each year providing the country with about 80% of its annual energy needs. The situation has changed drastically since then, however. Currently, coal mined in Belgium provides the country with only 14% of its annual energy requirements. Some 8 million metric tons of Belgian coal are used there each year. The remaining 86% of Belgium's energy needs is met by imported natural gas, petroleum products, and by coal imported from elsewhere. The chief reason for the change is the escalating costs of mining coal.

The first gasification test probably will be conducted at the Borenaga coal field in southern Belgium near the French border. Air, at pressures ranging from 20 to 50 atm, will be forced down through a pipe extending into the coal seams 1,000 m or so underground. Electrical heaters will be used to start combustion. The resulting gas, a mixture of carbon monoxide, hydrogen, and nitrogen, will come to the surface through neighboring pipes. The gas will have a low calorific value, probably between 100 and 140 Btu per cu ft. But it may be possible to harness it to generate electricity both by raising steam in heat exchangers and by passing it through gas turbines.

The combination of the steam turbine and gas turbine cycle to generate electricity has a high degree of efficiency, upward of 40%. Allowing for some loss due to energy consumption in the process, net electricity generating efficiency stands at about 36%. It probably will be possible to utilize about 50% of the calorific value of the underground coal. A further 13% or so efficiency could be gained from the high temperature of the emerging gas, probably on the order of 700°C, to bring total utilizable energy to about 63%. Multiplying this value by the 36% net efficiency factor of the combined cycle works out at 22.7% of the original energy of the coal field that could be converted to electricity.

Exploratory drilling is to begin in 1977. Underground burning tests probably won't get started until early in 1978. All told, the project will take about three years and cost about $13 million. However, it is open to other participating countries that may wish to join in.

## SOVIET EFFORT

### Background

A review of the Soviet work in underground gasification of coal clearly shows that the Soviet effort exceeds the combined efforts of all nations. The Soviet decision to pursue underground gasification of coal was made in 1928 with the first field experiments being performed in 1933. Their work has continued up to the present, with varying levels of effort that appeared to peak in the late 1960s. It is indicated that as many as 3,000 people were working on underground gasification in 1963. It appears that the effort might have more than doubled by the late 1960s and then fallen off dramatically by the early 1970s. The reduction was probably due to the Russian discovery of large resources of natural gas, and the construction of the necessary pipelines for its distribution.

A crude economic estimate, taking into consideration the numerous commercial plants they constructed and operated as well as the number of people involved over the past years, indicates that it might take as much as ten billion dollars at today's prices to reproduce their work. Since the USSR operated numerous commercial power plants over long periods of time, it is important to understand not only what their final process designs looked like, but also why they settled on such designs. It is easy to conceive of numerous designs for the underground gasification of coal that address one particular problem or another, but in the operation of a successful commercial process all the problems have to be dealt with simultaneously in a satisfactory and harmonious manner. The Soviets are the only ones who have demonstrated that they have a system design that can be operated repeatedly in a predictable manner, not only in the same area, but can be transferred from one area to another where there are large differences in the geological nature of the coal seams.

### Process Designs

The purpose of underground gasification of coal is to convert coal into a combustible gas by carrying out the appropriate chemical process underground. The Soviets saw this as a possible means of recovering the heating value of the coal without having to employ men in the dangerous and unhealthy job of underground mining. Air, and sometimes oxygen-enriched air, is used to partially combust the coal to produce a combustible gas. In review, the essential reactions describing the process are:

(1) Coal + heat $\rightarrow$ char (carbon) + $CH_4$ + $H_2$ + tars + $CO_2$ + CO + etc. (pyrolysis)
(2) Char + $O_2$ $\rightarrow$ CO + $CO_2$ + energy
(3) Char + $H_2O$ $\rightarrow$ CO + $H_2$ - energy

It is important to add water to use the excess energy released in reaction (2) to produce a combustible product via reaction (3). In practice, the Soviets found that generally the amount of water intruding into the gasification zone naturally, was more than sufficient for air gasification, so no attempt was made to add water to the injected gas.

The Stream Method for gasifying coal in steeply dipping coal beds was the first design that the Soviets felt had promise (see Figure 3.5 p 40). They tried many schemes in the first few years of their field effort with the intent of transferring surface gasification technology underground. The first attempts involved men underground, using explosive fracturing (the Chamber Method), and excavating or drilling holes from one drift to another (the Borehole-Producer Method). Poor economics and unstable operation led to the abandonment of these methods.

The Stream Method was first tested at Lisichansk in 1935; in the design the injection and exhaust holes were drilled along the coal seam and were connected at the bottom by a mined shaft. The flame was initiated in the connecting channel and gradually spread over the entire length. The flow had to be reversed periodically in order to approximate a "horizontal" burn front that moved up the seam. The key feature of this system was that as the coal was consumed, more coal would fall into the void that was created. Thus, the coal was automatically rubblized and fed into the combustion zone. This feature was retained in all further designs for both steeply dipping and horizontal coal beds. It might even be considered as a fundamental design requirement for any underground coal gasification system.

There appeared to be two major problems with the Stream Method. These were resolved in later designs by the modifications shown in Figure 9.1. There was a tendency for oxygen (air) to leak through the coal between the two access holes, above the region where the combustion was taking place. This would either consume part of the product gas, or would introduce a hazardous level of oxygen in the product gas. Also, there was a tendency for the combustion front to channel over the top of the coal, resulting in reduced resource recovery and erratic gas quality. These problems were resolved by drilling and casing the air injection holes through the overburden. In this manner the injection holes were isolated from the exhaust holes. Also, the probability of the flame front channeling over sections of coal was minimized by fixing the injection point at the bottom. If the coal seam was thicker than approximately 8 m, it was necessary to drill the injection holes at a slant "under" the seam in order to avoid shearing the wells off by subsidence. However, when these expensive slant holes were used, vertical holes were usually also drilled to help establish the initial bottom manifold by hydraulic fracturing or reverse combustion. The vertical holes were also used in the first stages of gasification.

## FIGURE 9.1: TYPICAL SCHEME OF AN UNDERGROUND GAS GENERATOR FOR STEEPLY DIPPING COAL BEDS

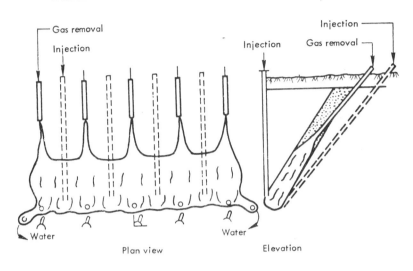

Plan view          Elevation

Source: UCRL-50026-75-4

Data show that the heating value always increases with increased gasification intensity and reduced water-inflow rate. In theory there is an optimum water intrusion rate for any given injection rate. In practice, this was unimportant to the Soviets since they usually had higher water intrusion rates than desirable, forcing them to operate on the water rich side of the optimum.

Relationships can be determined which make it possible to establish a permissible flooding level for the coal seam being gasified, and correspondingly the amount of drainage required. The rate of gasification, as given by the number of tons of coal gasified per hour, does materially affect the moisture content. This relationship is of utmost practical importance, because it gives a means of directly controlling the calorific value of the gas.

When coal seams of different thickness are gasified, the heat of combustion of the gas produced from the various seams varies widely. In general, the calorific value of the gas decreases with thickness, especially in the range from 0-2 m thick. For example in order to obtain a calorific value of 1,000 kcal/m³ in a seam 1 to 2 m thick, the moisture content must be equal to or less than 120 and 200 g/m³, respectively. In actual practice, in the Kuzbass generators the water content in these seams is typically 250 and 400 g/m³. Therefore to obtain a gas with calorific value of 1,000 kcal/m³ from the Kuzbass or Donbass coals, the seam must be thoroughly drained. Also, the gasification rate must be increased so that the water inflow rate is below 0.6 and 1.0 m³/ton of coal, respectively. Furthermore, the heat loss to the surrounding rocks increases as the seam thickness decreases. This heat loss results in a further lowering of the heat of combustion of the gas.

The Soviets, therefore, conclude that owing to the low calorific value of the gas, the high gasification intensity required, and the extensive drainage required, bituminous coal seams of 1 m or less thickness are not economical to gasify. Lignite seams of 2 m thickness or less are also believed to be uneconomical to gasify (1). These considerations go a long way in explaining low gas quality produced in early trials in Britain and the United States (2)(3).

The 1- to 2-m limitation in coal seam thickness has very important implications. Most of the coal in the eastern part of the United States lies in seams less than 2 m thick, while most of the coal in the western United States is in thicker seams.

### Design Study

Figure 9.2 shows a typical plan view of the Soviet process for horizontal seams. The dotted lines are meant to show the location of the underground linkage channels formed in the coal along the bottom of the seam by a countercurrent combustion step in preparation for gasification. The production phase of gasification is carried out by cocurrent combustion in the channels. Cocurrent and countercurrent refer to flame front propagation in the same direction as the gas flow or in the opposite direction, respectively.

Figure 9.3 illustrates how the drilling pattern and operational method were modified to accommodate thick coal seams. The primary problem to be solved was that with thick seams, the ground subsidence had a tendency to shear off the access pipes prematurely. To resolve this, two modifications were made: [1] Both the injection and production pipes were

## FIGURE 9.2: PLAN VIEW OF THE SOVIET PROCESS FOR HORIZONTAL SEAMS

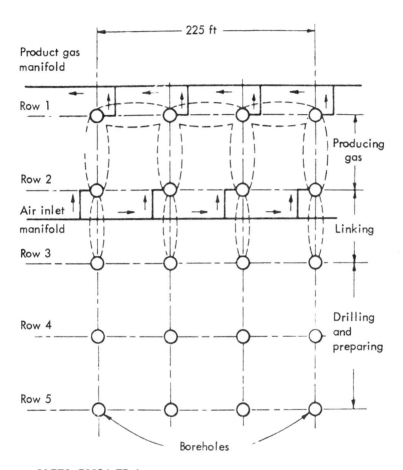

Source: UCRL-50026-75-4

drilled at a slant so that they were out of the subsiding region until after they were no longer needed; [2] A lateral withdrawal or injection technique was invoked. A network of wells was set up such that, in plan view, rows of injection wells were normal to that of the production wells. When gasification is carried out in this manner it can be seen that the row of pipes along the side of the zone being gasified is never affected by subsidence, and the pipes in the pattern directly over the coal being gasified are not affected until after they are no longer needed. Therefore, with this design one is never attempting to operate with a pipe that has been exposed to any ground movement due to subsidence.

## FIGURE 9.3: MODIFICATIONS FOR ACCOMMODATING THICK COAL SEAMS

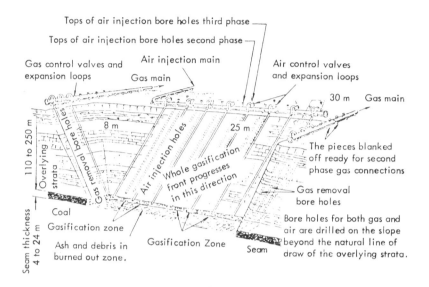

Source: UCRL-50026-75-4

The Soviet underground coal gasification systems for horizontal and steeply dipping beds, which are essentially the same basic design at different angles to the surface, are deceptively simple. However, it has been demonstrated repeatedly that they can be made to operate successfully in an extremely complex environment, a burning coal seam with a subsiding roof. The basic system design can also be transferred from one coal seam to another with reasonably predictable results, in spite of the wide variations in the geological conditions that exist between coal seams.

**Design Characteristics**

It is useful to understand the design in terms of how it handles a number of easily recognized physical phenomena that must be dealt with in any design for the underground gasification of coal.

Gas Leakage: One of the primary design requirements of an underground coal gasification process is that it minimize the loss of both the injected blast and the product gas, which can leak through cracks to the surface or other surrounding formations. Since the extent of subsidence-caused leakage cracks is unpredictable as the combustion zone moves across a coal seam, it is important to design the system so that it has minimum sensitivity to such leaks. The only way this can be done is to be certain that the permeability in the coal between the injection and exhaust holes is always much higher than the permeability from the reaction zone to the surface through subsidence. The Soviets accomplished this by establishing very highly permeable paths in the coal (linking) between the injection and exhaust holes before gasification and then operated at the lowest possible consistent with production requirements. They also helped seal the cracks at the surface by filling them with mud using a bulldozer.

High Gas-Flow Rate at Low Pressure: A high gas-flow rate is necessary to compensate for the water intrusion rate as well as to give an economically high production rate. This requirement is tightly coupled to the gas leakage problem. (Increased flow rates require increased pressure drops.) The high gas-flow rates can be maintained with minimum leakage only if the permeability between the injection and production holes, exceeds some minimum value.

Directional Control of Gas Flow: Directional control of gas flow is essential for a reproducible system. The construction of highly permeable linkage paths at the bottom of the coal seams, with predetermined spacings, give the reproducibility in flow control that is needed. It makes the process insensitive to variations in the natural coal permeability and thus enables one to design a predictable system after obtaining initial scaling data.

High Surface Area Reactor (Packed Bed) with High Permeability: The high permeability is needed for the reasons stated above. The high surface area, which infers a rubblized zone of coal without bypass channels, is almost always needed for any efficient gas-solid reaction. In such reactions, the reaction rate is limited by the surface area of the solid, since this is where the initial reactions must take place. The Soviet system for horizontal coal seams accomplishes this rubble zone at the tip of each channel by the flame front undercutting the coal and the overlying coal falling into the void. The system for steeply dipping beds accomplishes this by coal falling into the void created at the bottom where combustion is initiated. It is also expected that the reaction rate is enhanced at the coal surfaces in the upstream region of the channel, where the channel width is many times the thickness of the coal seam, by thermal spalling of the coal face.

Liquid Control: One of the primary effects that liquids (water and pyrolysis bars) can have on the propagation of the flame front in a horizontal seam is to cause the flame front to move preferentially along the top of the coal seam. This can result in very poor resource utilization. The Soviet process minimizes this effect by initially forming channels at the bottom of the seam in the preparatory linking step. The channels are kept quite hot all during the process so that any liquids that intrude into the region, as well as liquids formed by pyrolysis, are carried out as vapors.

Maximized Survival and Spacing of Access Pipes: A principal portion of the cost of carrying out underground coal gasification is associated with the four access wells. It is thus important to maximize the probability of their survival during the process and of the removal and reuse of the casing. It is also important to maximize the coal removed per hole, which dictates an optimum between fractional resource recovery and hole spacing. The Soviet designs remove the access holes from the subsiding zone until they are no longer needed. This both maximizes the probability of their survival and also makes it possible to pull the casing and reuse it.

A System That is Adaptable to Thick or Thin Seams: The Soviets have found no maximum limit in thickness of coal seam for their processes. They have operated in seams up to 60 ft thick with no indication that they are approaching a limit. However, the Soviets did find that there was a limit on how thin a seam could be. If the seam thickness fell below 3 to 4 ft, the heating value of the gas became unuseably low. As the coal seam gets thinner, the fraction of the heating value of the coal that goes into heating the surrounding rock increases to a prohibitive level.

No Men Underground: The Soviet design requires no men underground. This is not true of the British design or the early Soviet designs. It is an extremely important feature since once gasification is initiated, it is almost impossible to guarantee that the toxic product gases would not leak into nearby mine workings.

A System Than Can Be Applied to Multiple Layered Coal Seams: The Soviets have been able to sequentially gasify multiple layers of coal starting with the top seam and working down.

A Continuous System As Well As An Intermediate Load System: A system that operates continuously (the Soviet system sweeps continuously across a coal seam) is advantageous. However, there is also a need for electricity on an intermittent basis for supplying intermittent loads. The Soviet system in some cases can be turned on and off in time periods of a few hours which make it useful for supplying intermittent loads.

Constant Gas Composition vs Time: Due to the high surface area, in the rubblized zone of coal for both the horizontal and steeply dipping bed systems, it is possible to maintain a product gas composition that is quite constant in its heating value. This requires one additional control in that the flow rate is varied to "fine tune" the heating value.

Reproducibility, Predictability and Control: The performance of the process design is reproducible and predictable within reasonable limits from one generator to another within the same coal seam as well as when it is transferred from one coal bearing area to another. It has been possible to make it operate in widely varying coal types, (lignite, subbituminous, and bituminous). It has been demonstrated to have adequate controls to achieve a high resource recovery and a constant preshot gas quality.

Minimum Sensitivity to Coal Swelling: The large dimension channels formed in the linking phase were not plugged by moderately swelling coals.

Minimum Sensitivity to Flame Front Channeling: The Soviet system makes no attempt to avoid the nominal tendency of flame front channeling, in fact it encourages it. However, the gas quality is insensitive to such channeling because the channels are made very long with a rubble zone at one end or the other. This can only be accomplished because gasification is carried out along the plane of the seam where it is possible to make the channels almost arbitrarily long.

Simplicity of System Design and Operation: The Soviet design is exceptionally simple (deceptively so) as is its operation. It involves only the technology of drilling a simple pattern of holes (although slant drilling is not always easy) and handling compressed air. The end result is that it is very insensitive to the many uncontrollable physical phenomena that operate in the process. Such simplicity and insensitivity are essential features that make the process work.

## UNDERGROUND GASIFICATION IN CZECHOSLOVAKIA

Extensive research on underground gasification of coal has been conducted in Czechoslovakia. In 1956 the first trials of underground gasification were performed in the northern Bohemia brown coal basin. From 1959 to 1964 the first gasification tests on low-grade brown coal were performed in Borislav, USSR. However, the test program was only small-scale, because of the limited coal reserves in Borislav. Tests on underground gasification of brown coal reserves were begun in the Chomutov region in 1964. After special equipment and installations had been set up, large-scale tests took place from 1964 to 1966. Gas yields of 20,000 Nm³/h were obtained.

## Geological Conditions

The test area lies in a 120 m thick tertiary complex, consisting of a roof series, the seam hopper, and a floor series. The upper seam had an average thickness of 1.4 m and the lower one a thickness of 1 to 1.5 m. Owing to the small thickness of the seam and the low quality of the coal, these seams were not considered for underground gasification.

The middle seam, used for gasification, was divided into two layers by an intermediate bed of carbonaceous clay 1.5 m thick. The upper bed was about 2.5 m thick and the lower one 2 m. About 109,000 tons of coal were available for gasification.

## Drilling

From 1960 to 1961 drilling took place mainly without rinsing. To avoid losses to the surrounding medium, the walls of the gas and air boreholes were cased with cement. After the cement had set the boreholes were sunk as far as the seam with a drill of smaller diameter before the start of the operation. The boreholes were drilled about 1 m deeper. The additional borehole lengths were not lined with cement and formed the active parts of the holes. The covering layers of the middle seam lay at depths between 33 and 45 m. Thus, the depth of the 8 rows of boreholes varied between 36 and 51 m.

From 1963 the seam, containing much water, was drained by means of drainage boreholes, so that the water inflow could be reduced from 47 to 2 liters/minute. The work was then completely stopped in May 1965, without a reduction in the gas quality being observed. The drainage work served mainly for research purposes. The performance of the individual pumps was investigated and the radius of action of the depression cone determined. The water inflow was below the critical values. Above ground the casing string was provided with a special head and fittings for air and gas regulation. The whole system of coal seams and boreholes is called the underground generator.

## Test Program

The test program took place in three stages, a particular seam section being gasified in each stage. The first stage went into operation on January 31, 1964 and ended on June 15, 1965. The test program of the second stage ran from June 16, 1965 until April 20, 1966. The test series of the last stage lasted from June 23, 1966 to December 4, 1966. This stage served to determine the necessary safety precautions for future introduction of underground gasification. For this reason the results of the tests were not evaluated further.

## Operation of the Generator

In the first stage about 38,900 tons of coal reserves of the upper bed were gasified. Since the coal in the upper bed was not completely gasified, the remaining coal together with the lower bed could be transformed into gas in the second stage.

The operation begins with ignition of the seam and the establishment of a connection between the individual boreholes. This connection is formed by fire channels, obtained by flowing air at a pressure of about 6 bars into an adjoining borehole. Air movement in all directions is produced in the seam, and thus also in the direction of the ignited borehole.

As soon as the fire channels are ready, actual gasification of the seam begins. The channels widen and a combustion zone is formed. For the gasification, abundant air at low pressure is blown in. During the gasification continuous sinking of the layer above the gasified seam takes place, with the formation of a subsidence trough. In Brezno lateral gas removal was chosen; in this variant the gas is only removed from special boreholes on the side of the generator.

## Aboveground Installation

Adjustable distribution networks for steam and water were attached to the generator. The air boreholes were provided with inlet pipes and regulating devices for high- and low-pressure air. From the central installation the total generator area was supplied with air by means of a stationary main pipeline system with adjustable connecting circuits. The diameter of the low-pressure pipes was 200 to 300 mm. The pressures occurring were up to 2.5 bars. In the main pressure conduits with diameters of 80 to 100 mm, the pressure was up to 6 bars. The air supply of the generator was maintained by a turbo-compressor (P = 6 bars, Q = 6,000 Nm³/h) and blower (P = 2.5 bars, Q = 1,500 Nm³/h).

The gas produced had a temperature of about 400°C and was collected in gas boreholes. From there it was led through insulated pipes 500 mm in diameter, after removal of the mechanical impurities, into two gas boilers installed on the surface. The boilers had a capacity of 10 tons of steam per hour. Although most of the gas extracted was burned in the crude state, stabilizers promoting combustion were added to a small amount of the gas.

Most of the time the combine had to be supplied with electricity from the national grid. The water came from the public supply via a cooling tank with water treatment in the boiler unit. The other ground-surface installations consisted of workshops, laboratories, industrial buildings, and washrooms.

## Results

Time and Safety: The combine operated without interruption from January 31, 1964 to April 20, 1966. There were a few stoppages due to a break in power production. With somewhat impermeable upper beds, and with the observation of the necessary safety regulations there is no direct danger to workers due to leakage from old shafts, because relatively large distances can be chosen.

Work began again on June 23, 1966 (this was the beginning of the third stage), but this time only on a limited scale and with direct removal of the gas. The third stage ended on December 15, 1966. It served only to determine the necessary safety precautions for a large generator and also to determine the possible gasification pressures. The possibilities of preventing undesirable gas losses were also to be investigated.

Evaluation: The following table compares the test results from Soviet underground gasification in brown coal fields with the results from underground gasification tests in Czechoslovakia.

### Comparison of Test Results on Underground Gasification in the Soviet Union and Brezno, Czechoslovakia

|  | Large-Scale Test in Brezno | Gasification Test in the Soviet Union |
|---|---|---|
| Breakthrough rate, m/d | 1.075-0.85 | 0.6-1.0 |
| Measured air consumption for breakthrough, $Nm^3$/d | 10,093-9,787 | 15,000-19,000 |
| Calorific value of the gas, kcal/$Nm^3$ | 917-911 | 800-850 |
| Gas loss, % | 24.1-19.9 | 20.7-28.5 |
| Chemical efficiency of the underground gasification, % | 69.5-71.6 | 60.0-70.0 |
| Seam recovery, % | 78.2-96.0 | 80-100 |

Results indicate that the second gasification stage of the whole remaining middle seam with an average thickness of 4.5 m of utilizable coal gave better results than the first stage. The occurrence of the main break-throughs between January 31, 1964, and October 14, 1964, in the first stage can be considered responsible for this effect. These occurred at a low average system capacity of 5,270 $Nm^3/h$ in the first stage compared with 16,865 $Nm^3/h$ in the second stage. The second stage, lasting from June 16, 1965, to April 20, 1966, is characteristic of normal operation of an underground generator. Evaluation of the data shows that under suitable geological conditions it is possible to utilize brown coal reserves which are not suitable for mining.

## REFERENCES

(1) R. I. Antonova et al., *Khim. Tverdogo Topl.*, No. 1, p. 86 (1967). [Lawrence Livermore Laboratory, Rept. *UCRL Translation 10804* (1974).]
(2) National Coal Board, Great Britain, *The Underground Gasification of Coal*, Pitman and Sons, London, (1964).
(3) J. P. Capp, R. W. Lowe, and E. F. House, *Underground Gasification of Coal; Operation of Multiple-Path System*, U.S. Bureau of Mines, Rept. R15830 (1961).

# Environmental Implications

The material in this chapter was excerpted from PB 209 274, UCRL-50026-75-3 and UCRL-50026-75-4. For a complete bibliography see p 251.

## GENERAL BENEFITS

Environmental hazards arise from the methods in which coal is mined conventionally, whether by open-pit or underground installations, and from the way coal is conventionally utilized.

Underground gasification produces a gas product which could readily be treated to remove particulates and its sulfur content. Environmentally, then, the treated gas would perform the same as natural gas. Coal itself, however, after combustion in furnaces retains residual problems of particulate and sulfur oxide emissions in stacks.

Waste dumps result from the coal-washing process and contain noncarbonaceous materials unwanted in the coal. These dumps can become quite large, unsightly and hazardous in populated areas. In underground gasification, such waste would remain underground. Moreover, the candidate underground gasification methods would need a filling material which would have to come from the surface. Old waste, fly ash, or other dumped materials, might be acceptable. On the other hand, if such materials are lacking, fill might have to be mined and processed, which could create other environmental problems.

Both conventional underground and surface coal mining methods cause substantial damage to the land. Disruptive damage from open pit and strip mining of near-surface coal seams is well known. An alternative to gasify such seams underground would eliminate the threat of unsightly abandoned strip pits and disruptions during mining and surface reclamation operations.

Underground gasification would eliminate conditions where uncontrollable and unwanted underground and waste-pile fires could occur. If the candidate methods are to be practical at all, subsurface combustion would be under positive control at all times.

The candidate methods for underground gasification should produce little or no effect on the hydrological character of the strata. The burned-out coal seam will have been filled with probably a low-permeable fill material. The combustion should have been total, leaving little sulfur in the ground. On the other hand, inducing high-temperature combustion conditions could cause heating of aquifers, otherwise unaffected, and an increase in dissolved mineral salts.

Because of emphasis on preventing roof collapse, no ground subsidence at all should be experienced with underground gasification, if successful filling techniques are developed. This contrasts with the practices in conventional underground mining, where roof collapse is sometimes deliberately induced in order to achieve high coal recovery. Detailed studies and analyses are included hereinafter.

Although underground coal gasification appears to possess clear-cut environmental advantages, when compared to conventional coal-recovery methods, there are potential environmental consequences that might become significant for a full-scale operation. Therefore, it is very important to take steps toward quantitatively evaluating the possible adverse environmental effects of underground coal gasification. LLL provided the required environmental assessment and arranged for monitoring activities and mitigation measures that will ensure compliance with environmental regulations.

The following discussion is the result of the study of the environmental implications as assessed from the Hoe Creek experiments performed by LLL.


## LLL HOE CREEK EXPERIMENTS

The following potential effects were among those given particular consideration in the environmental assessment of the Hoe Creek experiments:

1. Groundwater contamination and aquifer disturbance.
2. Air pollution from plant effluents or surface leaks.
3. Ground surface disturbances due to subsidence and explosive fracturing.

It seems clear that these possible consequences are generally insignificant for the Hoe Creek experiments because of the comparatively small scale of the operations, the isolation of the site and, in some cases, because standard methods of control and mitigation are applicable. Nevertheless a detailed consideration of these effects is essential.

Specal attention has been given to the question of possible groundwater contamination by reaction products that remain underground after the gasification process. These substances include the coal ash and the organic products of the combustion. Of particular concern are soluble compounds that may become dissolved in the groundwater as it percolates through the reaction zone following gasification. This leaching process might be expected to lead to groundwater contamination. Fortunately, however, site characterization measurements, together with laboratory investigations and computational studies suggest that the cleansing action of the surrounding coal will effectively restrict these potential contaminants to the immediate vicinity of the gasification zone.

It can be demonstrated that the coal matrix has a strong cleansing effect on the groundwater pollutants generated by the gasifier. In particular it was found that the high concentration of hydroxides (and metal ions) generated by the ash undergo reaction with bound phenolic end groups of the coal:

$$M^{+}OH^{-} + \underset{(bound)}{\underset{OH}{\bigcirc}} \rightleftharpoons \underset{(bound)}{\underset{O^{-}M^{+}}{\bigcirc}} + H_2O$$

Similarly the phenolic tars generated by the gasifier are strongly absorbed by the coal.

Computer code calculations of the movement of such pollutants into the groundwater demonstrate quite clearly that the pollution plume is retarded as a result of the cleansing effect of the coal.

Also considered are the possible effects of dewatering and underground subsidence on the groundwater system. The water produced during the pre-gasification dewatering operations is of sufficient quality for use by livestock and is to be piped to a nearby stock pond. Some contaminants may be present in the water produced during gasification, and special arrangements (reinjection or surface cleanup) can be made for its disposal.

Product gases from the production well are to be monitored and flared in compliance with ambient air standards. There is some possibility that subsidence-induced cracks will penetrate to the surface and result in gas leakage. To guard against the potential danger to operating personnel, hydrogen sulfide and carbon monoxide will be monitored at the surface during the gasification procedure.

By using a 10-m-high stack to flare the product gases the ground-level concentration of combustion products will be no more than $10^{-5}$ that of the stack concentration. Since the hydrogen sulfide concentration in the product gas is only $\sim 0.1\%$ by volume, then the resulting sulfur dioxide concentration (oxidized form of hydrogen sulfide) at ground level is less than 0.01 ppm.

Among the potential sources of surface disturbance are subsidence and explosive and explosive fracturing. A computational study of subsidence suggests that the surface subsidence will be, at most, roughly 2 ft, and that there will not be sharply defined edges. Seismic-damage estimates based on a variety of empirical data suggest that the explosive fracturing will not damage any significant man-made structures.

The question of possible uncontrolled burn has been carefully considered. Since the coal to be gasified lies well below the water table, and natural coal fires do not occur under these circumstances, it is felt that when oxygen and steam injection are stopped, and dewatering operations are discontinued, the burn will be effectively quenched.

From this it can be seen that the adverse environmental consequences of the Hoe Creek experiments appear to be small, and these experiments provide an ideal opportunity to assess the environmental effects associated with this approach in in situ coal gasification.

## CONTROL AND DISPOSAL OF POLLUTANTS

### Liquid Reinjection

The Felix No. 2 coal seam gasified in the Hoe Creek experiments lies well below the water table. This situation effectively precludes uncontrollable burn, but it also requires that the coal be dewatered before, and probably during, the gasification process. The water produced prior to gasification is entirely suitable for livestock use and is to be piped to a stock pond at a nearby ranch. However, the water that is pumped to the surface during the gasification process may contain contaminants, such as organic reaction products and inorganic materials leached from the coal ash. If such contaminants are present in significant quantities, the water is to be reinjected into the Felix coal formation through an existing well located about 300 ft from the gas production well.

The significance of the proposed reinjection is that the gasification zone itself will be the primary underground repository of reaction products and coal ash. Ground water that enters the gasification zone will dissolve some of the reaction products and leach inorganic materials from the coal ash. The dewatering and reinjection procedure consists of transferring some of this potentially contaminated water to the same underground formation approximately 300 ft away. In both places the contaminants are expected to be effectively confined to the immediate area by the cleansing action of the surrounding coal.

### Incineration of Product Gases

A second source of potential pollution is the mixture of gases produced in the gasification process. These product gases will include (in addition to the nitrogen from the injected air) hydrogen, carbon monoxide, carbon dioxide, methane, and about 0.1% hydrogen sulfide. To avoid air pollution and health hazards for operating personnel, these product gases are to be piped to an incinerator, which will convert and disperse the gases safely. The incinerator consists primarily of a cylinder approximately 5.5 ft in diameter and 15 ft high. A diesel-fueled auxiliary burner will provide pilot ignition and flame stability, and two flame detectors will initiate a warning signal if the flame goes out. The incinerator is designed to handle a gas flow rate of 1,500 scfm. Two blowers will produce ample air (3,000 scfm) for complete combustion.

The pipe leading from the production well to the incinerator will be insulated and heated to reduce condensation. However, some liquids are expected and these will be initially piped to a 1,000-gal "tar tank." The contents of the tar tank will subsequently be pumped into the incinerator and burned. In the event that complete combustion is not feasible, it is anticipated that permission will be sought for appropriate off-site disposal.

## Air Quality Calculations

Incineration of the product gases will yield, primarily, the combustion products water, carbon dioxide, and sulfur dioxide. In order to comply with ambient air quality regulations and avoid any possible adverse effects, it is necessary to ensure that the incinerator and exhaust stack will provide adequate dispersion and dilution of the combustion products.

A series of plume dispersion calculations have been carried out using a computer code developed at LLL. There calculations give the concentration of a potential gaseous pollutant as a function of distance downwind from a continuously emitting point source. Input to the calculations includes gas flow rates, exhaust stack geometry, local meteorological and topographic conditions, and effluent heat flux. The results of these calculations indicate, for example, that sulfur dioxide concentrations are very unlikely to exceed Wyoming's ambient air standards under the conditions of the projected experiment.

## Ambient Air Monitoring

A downwind air monitoring program is to be carried out during the gasification experiment. (This program is additional to the monitoring operations in the immediate area, which are primarily directed toward personnel safety.) The effluents to be monitored include: total suspended particulates, sulfur dioxide, hydrogen sulfide, and carbon monoxide.

Air sampling locations will be determined by daily weather conditions. Duplicate samplers will be placed downwind at two locations approximately 100 to 500 meters from the stack in an attempt to intercept the effluent plume. Sampling frequency will be determined by the results of initial daily sampling, and by changes in the gasification process.

## Emergency Episodes

Unexpected occurrences that might lead to the temporary release of product gases have been given special attention. An accidental release of these gases (which include carbon monoxide and hydrogen sulfide) would result in a potential danger to operating personnel but, otherwise, would not be expected to lead to significant adverse environmental effects.

Occurrences that would permit the release of product gases include: ground surface rupture (a possible result of underground subsidence), leaks in gas handling equipment, incinerator flame-out, and incomplete combustion within the incinerator. A prompt warning of any significant release of toxic gases will be provided by continuous air monitoring equipment, which will detect carbon monoxide and hydrogen sulfide. Effective methods for terminating any such gas release have been worked out. If necessary, personnel will utilize respiratory equipment while remedial action is underway. It should be noted that the underground gasification process can probably be stopped or significantly reduced in a few hours by shutting off the air injection system. If necessary, water could be injected into the gasification zone to quench the combustion process.

# Economics

The material in this chapter was excerpted from PB 209 274 and UCRL 75990. For a complete bibliography, see p 251.

## GENERAL BACKGROUND

A review of coal price trends over the post World War II period shows that there has been a major rise in the average price of bituminous coal. The rise in prices immediately after World War II largely reflected the inflation experienced at that time. While costs of labor and material rose steadily during the fifties and sixties, increases in productivity permitted the coal industry to survive despite a general softening in coal prices due to a shrinking market. In any event, the coal companies were not making much money.

In late 1968, however, a number of events set in motion a basic change in the coal industry. The Farmington mine accident accelerated the enactment of the Mine Health and Safety Act. The production of coal fell behind the growth in demand, particularly of the electric utilities; and the Japanese demand for metallurgical coals became significant in the marketplace. At the same time the rate of increase in productivity slackened, particularly in underground mines. All these factors helped to bring about the resulting price rise. Thus, compliance with the Mine Health and Safety Act, faltering productivity, and the surge in inflation were the major cost-push factors; while the entrance of the Japanese into the market, plus the apparent shortage of steam-coal (and of all fuels), were market-oriented factors that permitted the coal companies to pass on their rising costs and, in addition, improve their profit margins after many lean years.

Currently, the United States is finding itself facing an increasingly severe problem of finding new energy supplies to meet growing demand. With the almost daily variance of prices since the oil embargo, it is almost impossible to give actual projected costs of underground coal gasification versus other methods. The Lawrence Livermore Laboratory, in conjunction with its thick packed bed (TPB) studies, estimated costs in 1974 and these are presented here as a reference study.

## LLL COST ESTIMATE

Projected costs of pipeline quality gas produced by the LLL technique depend upon a number of factors. A parameter study was made using various depths and thicknesses of the coal bed, and two estimates for product gas composition (1).

The composition of the product gas can be bounded by two estimates. The optimistic estimate assumes that the products consist essentially of methane and carbon dioxide and that the sulfur in the coal remains trapped underground, while a more conservative estimate assumes that gas produced underground has the following composition:

| Product | Mol Percent (water-free basis) |
|---------|-------------------------------|
| $CH_4$  | 27.0 |
| $H_2$   | 19.2 |
| $CO_2$  | 20.4 |
| CO      | 32.4 |
| $H_2$   | 0.5 |
| $H_2S$  | 0.5 |

The above gas composition corresponds to the gases in equilibrium at about 1100° K (1520° F) and a pressure of 70 atm (1030 psi) and is somewhat higher in methane than the Lurgi gasifier products.

All the following estimates treat the more conservative case. All estimates assume annual production of 0.1 TCF of pipeline quality gas with 330 working days per year.

Estimated costs for the surface plant units using data from the El Paso Natural Gas Co. and Western Gasification Co. submission to the Federal Power Commission (2)(3) are shown in Tables 11.1 through 11.3.

## TABLE 11.1: INSTALLED PROCESS EQUIPMENT
## GAS PROCESSING PLANT

| Unit | Installed Cost (1974 $) M$ |
| --- | --- |
| Oxygen | 39.6 |
| Tar and dust removal | 0.2 |
| Shift conversion | 10.7 |
| Purification | 36.7 |
| Sulfur plant | 3.2 |
| Methanation, compression, drying | 20.9 |
| Total | 111.3 |

Source: UCRL-75990

## TABLE 11.2: TOTAL INVESTMENT FOR GAS-PROCESSING PLANT

| Item | Cost, M$ (1974 $) |
| --- | --- |
| Installed process equipment | 111.3 |
| Utilities and offsites | 62.5 |
| Indirect costs | 47.2 |
| Total | 221.0 |

Source: UCRL-75990

## TABLE 11.3: OPERATING COSTS FOR
## SURFACE PLANT FACILITIES

| Item | Cost, M$ (1974 $) |
| --- | --- |
| Coal for process heat | 2.67 |
| Catalyst and chemicals | 1.10 |
| Water | 0.72 |
| Maintenance materials and supplies | 4.82 |
| Personnel and consultants | 5.89 |
| Total | 15.20 |

Source:UCRL-75990

Costs for preparing the coal bed for gasification depend upon the depth and thickness of coal. An example is given in Table 11.4 for the Thunderbird site in the Powder River Basin near Gillette, Wyoming (4). Three coal beds with a net average of 244 ft thickness are found in a 600 ft thick coal/shale sequence and the average depth to the top of the upper coal bed is 1,185 ft. It is assumed that the entire 600 ft sequence must be fractured in order to gasify the coal.

### TABLE 11.4: OVERALL COSTS—CONSERVATIVE ESTIMATE*

| | Capital Investment, M$ (1974) |
|---|---|
| Surface plant | 221.0 |
| Drilling | 4.2 |
| Startup* | 10.2 |
| Subtotal | 235.4 |
| Working capital** | 5.9 |
| Total | 241.3 |

| | Operating Costs, M$ (1974) |
|---|---|
| Surface plant | 15.20 |
| Drilling | 13.30 |
| Explosives | 4.90 |
| Coal royalty | 1.54 |
| Total | 34.94 |

*25% of operating costs + 1.5 M$ drilling materials.
**2.5% of subtotal.

Source: UCRL-75990

Other cases of coal bed depth and thickness were calculated in a similar manner. (See Table 11.5 for summary). The utility financing method recommended by the Federal Power Commissions Task Force (5) was then employed to produce the cost estimates shown in Figure 11.1. Costs for the optimistic estimate (i.e. product gas sulfur-free and essentially a mixture of carbon dioxide and methane) are in all cases $0.21/10⁵ Btu less than for the more conservative case.

Estimated costs for the Lurgi process and new processes under development by the OCR-AGA Program are shown for comparison in Figure 11.1. These costs correspond to use of low cost strip-mined coal (about $3-4/ton) which is the most favorable case for surface coal gasification.

## TABLE 11.5: CAPITAL INVESTMENT AND ANNUAL OPERATING COSTS FOR IN SITU COAL GASIFICATION, M$
### 0.1 TCF of gas at 945 Btu/scf, 1974$

| Coal Depth (ft) | Coal Thickness — 50 ft — Capital Investment | 50 ft — Annual Operating Costs | 100 ft — Capital Investment | 100 ft — Annual Operating Costs | 200 ft — Capital Investment | 200 ft — Annual Operating Costs | 244 Coal in 600 ft Sequence — Capital Investment | 244 Coal in 600 ft Sequence — Annual Operating Costs |
|---|---|---|---|---|---|---|---|---|
| 1,000 | 239.7 | 28.67 | 239.1 | 26.2 | 238.7 | 24.71 | 239.6 | 28.52 |
| 1,185 | 246.1 | 42.45 | 240.7 | 32.42 | 239.6 | 28.27 | – | – |
| 1,500 | – | – | – | – | – | – | 241.3 | 34.94 |
| 2,000 | 257.6 | 63.77 | 246.9 | 44.74 | 240.8 | 32.84 | 242.9 | 39.05 |
| 2,500 | – | – | 255.0 | 59.22 | 244.7 | 40.56 | 247.2 | 47.55 |
| 3,000 | – | – | 264.5 | 76.81 | 249.8 | 50.04 | 252.3 | 57.33 |
| | – | – | 275.8 | 97.7 | 255.6 | 61.05 | 258.0 | 68.39 |

Source: UCRL-75990

**FIGURE 11.1: GAS SELLING PRICES EXPECTED FOR THE LLL PROCESS FOR COAL AT VARIOUS DEPTHS**

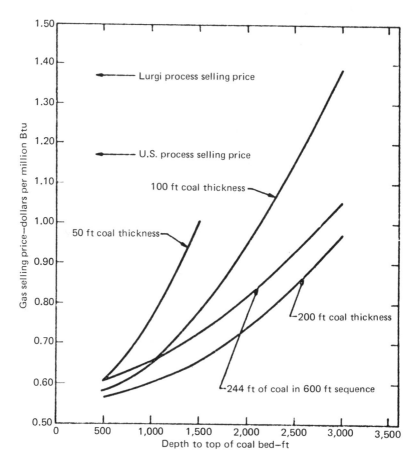

Coal thickness is a parameter. Gas selling prices shown for the Lurgi and new U.S. surface coal gasification processes are for strip-mineable western coal.

Source: UCRL-75990

It can be seen from Figure 11.1 that most of the coal bed thickness and depth models calculated would produce gas by in situ coal gasification at a cost advantage compared either to the Lurgi or the new U.S. surface techniques.

One existing coal deposit (the Thunderbird site) is represented by the case of 244 net ft of coal in a 600 ft coal and shale sequence with the top of the sequence at 1,185 ft. Over twenty billion tons of coal are estimated to be present in this deposit (4), which is equivalent to the proved reserves of natural gas in the U.S. if converted to pipeline quality gas. The cost estimate for pipeline quality gas from this deposit using the optimistic and conservative cost estimates, is bounded by $0.47-$0.68/$10^6$ Btu. There is a considerable potential cost advantage for in situ gasification using explosive fracturing.

Costs sensitivities to some parameters used in these estimates are given in Table 11.6. As can be seen, there is little incentive to attempt the more difficult task of using nuclear explosives for fracturing coal beds.

### TABLE 11.6: COST SENSITIVITIES TO VARIOUS PARAMETERS FOR THE THUNDERBIRD MODEL

| Parameter | Difference in Gas Cost Relative to Base Case (cents per million Btu) |
|---|---|
| Drilling costs, ± 100% | −7 to +14 |
| Explosive costs for strong or weak coal | −3 to +10 |
| High explosive costs, ± 100% | −3 to +5 |
| Nuclear explosives costs | −8 |

Source: UCRL-75990

The desired product in this study was assumed to be methane. If electricity is the desired product, then gasifying with air could be considered since nitrogen as a diluent should not pose a problem with a "mine-mouth" electrical plant. Electricity may be a desirable product with a load center nearby. However, based on current market trends, the location of most of the coal resource in the Rocky Mountain States, and the much higher cost of transmitting electricity than natural gas to population centers, it appears that pipeline gas via steam and oxygen gasification would be the preferred product by this technique.

For the cost estimates previously discussed, a large portion of the costs involved conversion of carbon monoxide and hydrogen to methane. These gases also can be used to make methanol, which may be preferred if methanol finds appreciable use as a transportation or boiler fuel (6).

## REFERENCES

(1) D. R. Stephens, *Revised Cost Estimate for the LLL In-Situ Coal Gasification Concept*, Lawrence Livermore Laboratory, Rept. UCRL-51578 (1974).

(2) Application El Paso Natural Gas Company before the U.S. Federal Power Commission, Docket No. CP73-131 (November 15, 1972).

(3) Application of Transwestern Coal Gasification Company, Pacific Coal Gasification Company, and Western Gasification Company before the U.S. Federal Power Commission, Docket No. CP73-211 (February 2, 1973).

(4) J. S. Mold and T. C. Woodward, "Project Thunderbird—A Nuclear Trigger for Coal Gasification," *Coal Age*, September, 1967, pp. 64-68.

(5) Synthetic Gas-Coal Task Force, The Supply-Technical Advisory Task Force—Synthetic Gas Coal, Federal Power Commission (1973).

(6) A. D. Pasternak, *Methyl-Alcohol Production by In-Situ Coal Gasification*, Lawrence Livermore Laboratory, Rept. UCRL-51600 (May 30, 1974).

# Future Trends

The material in this section is excerpted from CONF-751171-1 and PB 256 155. For a complete bibliography, see p 251.

## DEVELOPMENT IN THE U.S.

The expectations for underground coal gasification should no longer be based upon technical feasibility uncertainty, but rather upon how the relatively well defined products can be utilized. It has been established that the coal resource and the marketplace for underground coal gasification must be located close together. Since the products are useful for both power generation and chemical feedstocks, the closeness of refineries, chemical or petrochemical industries should be an important consideration in underground coal gasification site selection. The value of underground coal gasification products for process heat should also not be overlooked.

For the first time in U.S. history, it appears that the underground coal gasification process and products are now economically attractive and at least competitive with other synthetic gaseous fuel process alternatives. Therefore, this process needs to be demonstrated on a commercial basis in the U.S. to evaluate it as a means of providing a secure, safe and cheap source of power for the American consumer.

The factor today that is decidedly new and did not exist when experimental work ceased in the early 1960s is the ability to undertake remote sensing by numerous techniques, to manage the enormous amount of data collection by computer monitoring, and to process the data for a given purpose. In addition, the data can be displayed graphically and the display made to illustrate pictorially the situation underground. Several planned experiments will provide for the collection of data from remote sensing devices by computer. In the longer run, as more is learned about the factors that can control in situ gas quality and quantity, it may even be practical to build in feedback loops to assure better control over the gas-ification process.

The most important factor in furthering the development of underground coal gasification in the U.S. is for strong industry participation in all phases of R&D. One of the major technologies that need to be developed further in the U.S. to enhance the underground coal gasification process is directional drilling. The directional drilling-gasification schemes developed in the Soviet Union also greatly reduced the need for expensive in situ monitoring systems. That is, with the closely spaced, parallel, directionally drilled holes such problems as "fingering" or loss of control of the symmetric radial burn mode from vertical wells is largely eliminated.

The feasibility of underground coal gasification technology has been adequately demonstrated. Utilization of the previously developed technology and methods could provide an expedient, short term means of recovery of our most abundant energy resource.

# Bibliography

The following reports used in the preparation of this book are available from:

National Technical Information Service
U.S. Department of Commerce
5285 Port Royal Road
Springfield, Virginia 22151

CONF 751171-1    L. Z. Shuck and J. Pasini III: "A Report on Progress in Underground Coal Gasification." The International Energy Engineering Congress, Chicago, Ill., Nov. 4–5, 1975, Session C-11.

MERC/SP-75/1    L. Z. Shuck and G. E. Fasching; *Instrumentation for In Situ Coal Gasification*, August 1975.

PB 209 274    *A Current Appraisal of Underground Coal Gasification*. Arthur D. Little, Inc. April 17, 1972.

PB 241 892    R. F. Chalken; *In Situ Combustion of Coal for Energy*. November 1974.

PB 241 925    G. G. Campbell, C. F. Brandenburg, and R. M. Boyd; *Preliminary Evaluation of Underground Coal Gasification at Hanna, Wyoming*. Oct. 1974.

PB 256 155    *Outlook for Research and Development in the Underground Gasification of Coal*. Arthur D. Little, Inc. Nov. 7. 1975.

SAND-75-0459          H. M. Stoller; *In Situ Instrumentation Applied to Underground Coal Gasification.* Sept. 1975.

SLA-73-0919          D. A. Northrop; *Instrumentation Systems Development for In Situ Processing.* Sept. 1973.

TID-26825          *The LLL In-Situ Coal-Gasification Research Program in Perspective*—A Joint *GR&DC/LLL Report.* June 10, 1975. Contributers are:

| GR&DC | LLL |
|---|---|
| R. L. Arscott | C. R. Adelmann |
| J. C. Fair | J. L. Cramer |
| A. M. Garon | D. O. Emerson |
| R. Raimondi | A. Maimoni |
| R. P. Trump | A. D. Pasternak |
| D. L. Bidlack | D. R. Stephens |
| | D. S. Thompson |

TID-27023-P2          R. C. Rupert, R. Choate, S. Cohen, A. A. Lee, and J. Lent; *Energy Extraction from Coal In Situ: A Five-Year Plan. Volume II: Technical and Management.* Sept. 21, 1975.

TID-3349-05345          L. A. Schrider and J. Pasini; "Underground Gasification of Coal, Pilot Test, Hanna Wyoming." AGA meeting Chicago. Oct. 29-31, 1973.

TID/LERC-7          L. A. Schrider, C. F. Brandenburg, D. D. Fischer, R. M. Boyd and G. G. Campbell; *Outlook for Underground Coal Gasification.* May 1975.

UCID-16155          A. E. Sherwood. *Thermal Wave Propagation Models for In Situ Coal Gasification.* October 27, 1972.

UCID-16631          R. H. Cornell; *A Methodical Approach to Temperature and Pressure Measurements for In Situ Energy—Recovery Processes.* Nov. 14, 1974.

UCID-16640          T. J. Burgess, P. A. Schultz, and L. F. Wouters; *Optical Remote Measurement of Temperature in the Range 0-2000°C.* Nov. 8, 1974.

UCID-16817          J. H. Campbell; *Preliminary Modeling of Roof Collapse and Calculation of Gas Loss, Water Influx, and Surface Subsidence Associated with the Packed-Bed Scheme of In Situ Coal Gasification.* April 22, 1975.

UCID-17007          C. B. Thorsness; *Estimates of Thermal Front Movements and Pressure-Drop-vs-Flow-Rate Relations in Forward In Situ Coal Gasification.* January 19, 1976.

UCRL-50026-75-1     W. Mead and H. L. Lentzner; *LLL In Situ Coal Gasification Program—Quarterly Progress Report January through March 1975.* May 26, 1975.

UCRL-50026-75-2     D. R. Stephens and H. L. Lentzner; *LLL In Situ Coal Gasification Program—Quarterly Progress Report April through June 1975.* Sept. 22, 1975.

UCRL-50026-75-3     D. R. Stephens and C. R. Schneider; *LLL In Situ Coal Gasification Program—Quarterly Progress Report July through September 1975.* Oct. 31, 1975.

UCRL-50026-75-4     D. R. Stephens and H. L. Lentzner; *LLL In Situ Coal Gasification Program—Quarterly Progress Report October through December 1975.* Jan. 30, 1976.

UCRL-51217 (Rev. 1) G. H. Higgins; *A New Concept for In Situ Coal Gasification.* Sept. 27, 1972.

UCRL-51676          D. W. Gregg; *A New In-Situ Coal Gasification Process that Compensates for Flame-Front Channeling, Resulting in 100% Resource Utilization.* Oct. 15, 1974.

UCRL-51686          D. W. Gregg; *Liquid Plugging in In Situ Coal Gasification Processes.* Oct. 25, 1974.

UCRL-51770          D. R. Stephens; *Prospects for In Situ Coal Liquefaction.* May 7, 1975.

| UCRL-51790 | J. R. Hearst; *Fractures Induced by a Contained Explosion in Kemmerer Coal.* April 15, 1975. |
| UCRL-51835 | R. L. Wong; *Coal Plasticity and the Physics of Swelling as Related to In Situ Gasification.* June 2, 1975. |
| UCRL-75990 | D. R. Stephens, A. Pasternak, and A. Maimoni; *The LLL In Situ Coal Gasification Program.* Aug. 16, 1974. |
| UCRL-76496 | D. W. Gregg; *Critical Parameters of In Situ Coal Gasification.* February 5, 1975. |
| UCRL-Trans-10854 | F. Jansen. Translated from German Patent 1,015,181. |
| UCRL-Trans-10855 | J. Rolfes. Translated from German Patent 952,837. |
| UCRL-Trans-10856 | J. Rolfes. Translated from German Patent 1,015,982. |
| UCRL-Trans-10866 | **H. Goerger and K. Engin. Translated from** "Verfahren der Untertagevergasung." *Glückauf-Forschungshefte* vol. 34, No. 5, 1973. |
| UCRL-Trans-10867 | H. Goergen and K. Engin. Translated from "Die Untertagevergasung in Brezno." *Glückauf-Forschungshefte* vol. 35, No. 1, 1974. |

The following U.S. Patents used in the preparation of this book are available from:

Commissioner of Patents and Trademarks
Washington, D.C. 20231

| U.S. Patent 3,497,335 | J. W. Taylor; Feb. 24, 1970 |
| U.S. Patent 3,628,929 | E. D. Glass and V. W. Rhoades; Dec. 21, 1971; assigned to Cities Service Oil Company |
| U.S. Patent 3,734,184 | J. O. Scott; May 22, 1973; assigned to Cities Service Oil Company |

U.S. Patent 3,770,398   G. E. Abraham and C. M. Royo; Nov. 6, 1973; assigned to Cities Service Oil Company

U.S. Patent 3,775,073   V. R. Rhoades; Nov. 27, 1973; assigned to Cities Service Oil Company

U.S. Patent 3,997,005   C. A. Komar; December 14, 1976; assigned to United States Energy Research and Development Administration

# COAL CONVERSION TECHNOLOGY 1976

### by I. Howard-Smith and G. J. Werner

*Chemical Process Technology Review No. 66*

The industrialized countries of the world are showing renewed and increasing activity in the area of synthetic fuels production from coal. Existing commercial coal conversion processes, such as the Fischer-Tropsch synthesis, the destructive distillation of coal, methanation etc. resulting in liquid and gaseous fuels, are again becoming economically competitive with petroleum.

The many processes and techniques of coal conversion have as a basic concept the transmutation of coal into forms acceptable to our transportation and heating equipments. To accomplish this, high-sulfur coals must be desulfurized, high-ash coals must be demineralized, and, most important of all, solid coal must be depolymerized into liquid and gaseous products that can be ignited and burned with facility.

Over 100 processes, in various stages of development, are available to carry out these procedures and are discussed fully in this book.

Originally prepared as an "in-house" report for the Millmerran Coal Pty. Ltd. of Brisbane, Australia, it is intended to provide readily accessible and concise data on all major activities in coal conversion technology. A special feature are the flow charts which clearly illustrate all the principal processes. A partial and very condensed table of contents follows here. Chapter headings are given, and some of the more important subtitles are also included.

ISBN 0-8155-0614-7

**133 pages**

# DEEP COAL MINING
## Waste Disposal Technology 1976

### by William S. Doyle

*Pollution Technology Review No. 28*

The environmental impact of coal mining has been recognized and described since medieval times. Drainage from active and abandoned mines has always been a source of river pollution, although former generations could afford to ignore it.

Coal refuse banks also are a source of acid mine drainage (AMD) and of silt. In addition these unsightly refuse banks are susceptible to spontaneous combustion and can smolder for months or years, thus contributing substantially to air pollution.

This book, based on government reports and important recent U.S. patents, discusses methods to prevent and control pollution associated with deep mining of coal. Various processes for neutralization of AMD, the role of iron oxidation and methods for oxidizing the iron are described. Treatments of AMD with soil or by ion exchange and reverse osmosis are detailed. One chapter describes the treatment of AMD combined with municipal wastewater.

Also discussed is the underground disposal of mine wastes, the necessary sealing compounds, as well as the extinguishing of burning refuse banks and their reclamation with the aid of planned vegetation.

A partial and condensed table of contents follows here.

Chapter headings are given and some of the more important subtitles are included.

ISBN 0-8155-0619-8

**392 pages**

# STRIP MINING OF COAL
## Environmental Solutions 1976

### by William S. Doyle

*Pollution Technology Review No. 27*

Late in 1974 the Office of Research & Development of the U.S. Environmental Protection Agency made the statistically well founded projection that for the remainder of the 20th century surface-mined coal will have to account for over 50% of our nation's production of this fuel.

Strip mining can be done responsibly without permanent damage to land and water. Technology exists for effective reclamation of mined lands, and such reclamation is being performed in some areas. Authorities emphasize the importance of planned pre-mining (fully discussed in this book) and point out that reclamation is less costly and more effective when integrated with the mining operation.

This book, based on 19 government reports issued from 1967 through 1974, describes surface mining of coal, land use and methods, land reclamation technology plus sediment and erosion control. Acid mine drainage, its sources, prevention and correction, as well as the mechanism of reclaiming acid strip mine lakes are discussed. Specific studies on revegetation, use of spoil amendments etc. are included. A whole chapter is devoted to West Germany's approach to the problem of strip-mined lands. The final two chapters are devoted to costs, economics, and financing.

A partial and condensed table of contents follows here. Chapter headings are given, followed by examples of important subtitles.

ISBN 0-8155-0611-2

352 pages

# HOW TO SAVE ENERGY AND CUT COSTS IN EXISTING INDUSTRIAL AND COMMERCIAL BUILDINGS 1976

## An Energy Conservation Manual

### by Fred S. Dubin, Harold L. Mindell and Selwyn Bloome

*Energy Technology Review No. 10*

This manual offers guidelines for an organized approach toward conserving energy through more efficient utilization and the concomitant reduction of losses and waste.

The current tight supply of fuels and energy is unprecedented in the U.S.A. and other countries, and this situation is expected to continue for many years. Never before has there been as pressing a need for the efficient use of fuels and energy in all forms.

Most of the energy savings will result from planned systematic identification of, and action on, conservation opportunities.

**Part I** of this manual is directed primarily to owners, occupants, and operators of buildings. It identifies a wide range of opportunities and options to save energy and operating costs through proper operation and maintenance. It also includes minor modifications to the building and mechanical and electrical systems which can be carried out promptly with little, if any, investment costs.

**Part II** is intended for engineers, architects, and skilled building operators who are responsible for analyzing, devising, and implementing comprehensive energy conservation programs. Such programs involve additional and more complex measures than those in **Part I**. The investment is usually recovered through demonstrably lower operating expenses and much greater energy savings.

A partial and much condensed table of contents follows here:

Much of the technology required to achieve energy savings is already available. Current research is providing refinements and evaluating new techniques that can help to curb the waste inherent in yesteryear's designs. The principal need is to get the available technology, described here, into widespread use.

ISBN 0-8155-0638-4

**725 pages**